Drip and Sprinkler Irrigation

2nd Fully Revised and Enlarged Edition

NIPA® GENX ELECTRONIC RESOURCES & SOLUTIONS P. LTD.

New Delhi-110 034

About the Author

Prof. R.K. Biswas is an M.Tech and Ph.D in Agricultural Engineering with specialization in Soil & Water Conservation Engineering. He has been actively engaged in teaching and research since he joined in Bidhan Chandra Krishi Viswavidyalaya in 1984. During this period Prof. Biswas has uninterruptedly undertaken the courses of irrigation, drainage & water management in PG & UG level and conducted research works funded by ICAR, MoWR, Govt. of India and other financial institutions. The outcome of the research works has been appreciated by the funding agencies and the professionals.

Prof. Biswas has served the State Govt. Departments (West Bengal) as the technical expert, Ex-officio director & TSG in the area of drought & water resources management, agro-industries and micro-irrigation development respectively. Prof. Biswas has been all along worked for the academic development of the institution and betterment of the students. The effort of Prof. Biswas in introducing UG & PG programs in Bidhan Chandra Krishi Viswavidyalaya is well recognized. Prof. Biswas has served the university at various capacities, viz. Dean, Post Graduate Studies; Dean, Faculty of Agricultural Engineering and Head, Dept. of Soil & Water Engineering.

Drip and Sprinkler Irrigation

2nd Fully Revised and Enlarged Edition

R.K. Biswas

Dean, Faculty of Agril. Engg
Member, Water Resource and Drought Mission (GoWB)
BCKV, Mohanpur, West Bengal

NIPA® GENX ELECTRONIC RESOURCES & SOLUTIONS P. LTD.
New Delhi-110 034

NIPA® GENX ELECTRONIC RESOURCES & SOLUTIONS P. LTD.

101,103, Vikas Surya Plaza, CU Block
L.S.C. Market, Pitam Pura, New Delhi-110 034
Ph : +91-11-43860225, Mob.: +91 9717133558, 9540816132
E-mail: newindiapublishingagency@gmail.com
Website: www.nipaersources.com

Print ISBN: 978-93-58876-55-0
ebook ISBN: 978-93-58878-05-9

Composed and Designed by NIPA®.

Preface to the First Edition

Water is become the precious commodity. The demand for irrigation water has increased many times over the years to bring more and more area under irrigation and simultaneously for increasing cropping intensity and to provide food and fiber to our ever-increasing population. Bringing more area under irrigation without the improved management may lead to stress to water resources and average success in production and productivity. Micro irrigation refers the drip and sprinkler irrigation and so far the most efficient method of irrigation. Micro irrigation has many advantages over the conventional methods of irrigation. However, the method needs some knowledge and skill for appropriate application of it.

This book **Drip and Sprinkler Irrigation** is intended as a text book of micro irrigation design and practices for the students of agricultural sciences and the professionals and workers in the field of micro irrigation. The book contains good numbers of numerical as example and task to get the students familiar to the requirements, complicacies, and possible remedies in actual working condition in addition to conventional broad and short questions. In every chapter of the book there are multiple choice questions to assist the students in attending competitive examinations.

The author is thankful to Indian National Committee for Irrigation and Drainage (INCID), Ministry of Water Resources, Govt. of India for financial support in conducting the research work, the outcome of which that are found relevant have been included in this book. The author gratefully acknowledges the support of Bidhan Chandra Krishi Viswavidyalaya and my esteemed colleagues who have all along inspired me to go ahead. I would like to acknowledge and appreciate the consistent persuasion of my students excepting which the things could not be materialized at all. The author acknowledges his indebtedness to the authors and researchers of various books, bulletins, monographs and published and unpublished papers from which most of the content of this book is drawn. Every effort is made to acknowledge the sources of information in the text. If any omission remains, it is inadvertent, and will be corrected if noted or pointed out.

I would like to thank NIPA for the support of publication of this book.

R. K. Biswas

Preface to the Second Edition

It is my pleasure to say that this book 'Drip & Sprinkler Irrigation' has been found useful to the students, professionals and many other users. Some sort of errors was identified in its first edition. Necessary cares have been taken to omit these errors. I may hope that this book will serve the purpose of the users for understanding the design, implementation and maintenance of micro irrigation projects.

I would like to thanks and acknowledge New India Publishing Agency (NIPA), New Delhi for the consistent support and persuasion in publishing this second edition.

R. K. Biswas

Contents

1

Drip Introduction

Introduction

Drip irrigation is an efficient method of application of water at the plant bottom at a rate nearly equal to the consumptive use rate of the plant, thereby minimizing the conventional water losses like percolation, runoff and evaporation from soil. It is a process of slow application of water on, above or beneath the soil by the surface, sub-surface, bubbler, and spray or pulse system. Fertilizer can also be applied with the drip water. Emitters or applicators are placed closed to the plants and used to spray water in the form of drops, tiny streams or miniature spray. In the drip system water applied from the point source advances in all direction in the soil outward from the source. Drip irrigation is essentially a low rate, low pressure, frequent and long duration application of water in plants root zone area.

Drip irrigation is also called as localized irrigation, trickle irrigation, daily flow irrigation, diurnal irrigation, drop irrigation, sip irrigation, and micro-irrigation. A particular name is being popularized in any area depending on the choice of the people of that area. The International Commission for Irrigation and Drainage (ICID) has recommended the term micro-irrigation while the American Society of Agricultural Engineers (ASAE) has preferred drip irrigation. In India it is told as drip irrigation.

Histories and Development

International

The surface or gravity and sprinkler irrigation can be traced back to 6000 B.C. along with the oldest civilization along the Nile, Tigress, Euphrates, Indus and Yellow river. Drip irrigation is considered a new approach and was developed originally as sub-irrigation. In 1860, experiments were conducted in Germany with the clay pipes with open joints laid below the surface of the soils where the basic idea underlying the drip irrigation could have been traced. In 1920, perforated pipes replaced these. Some activities were traced in Coloroda during 1913. In 1930, the irrigation system with 5cm galvanized iron pipe with triangular holes was developed in Australia.

In the early 1940's, an Israeli Engineer namely Symcha Blass observed that a big tree near a leaky tap showing more growth than other trees in the area, which were not richer with water from the tap. This observation led him to the concept of an irrigation system, which would have very small quantity of water, literally drop by drop. Eventually he designed and patented a low-pressure irrigation system. Drip irrigation was started in Denmark due to large-scale introduction of plastic pipes after the Second World War. The practice was adopted later in England. In early 1960's, drip irrigation showed great success in the desert areas of Negev and Arava in Israel. It was later introduced to USA and gained immediate and wider acceptance.

During 1969, the modern drip irrigation system began to be sold outside Israel on commercial basis. These were installed widely in USA, Australia and Mexico along with Israel and in smaller extent in Canada, Cyprus, France, Iran, New Zealand, United Kingdom, Greece and India. The area of drip irrigation increased to 54, 600ha in 1975 from mere 40ha in 1960. It further increased to 412760ha in 1981 and 1784846ha in 1991. The world wide survey conducted by the International Committee for Irrigation and Drainage (ICID) Group on Micro Irrigation in 1991 reported 35 countries in the world drip irrigation is being practiced. The highest area of 606, 000ha was in USA and the lowest 30ha in Ecuador. The India was in the seventh position with 70, 859ha.

Indian

Drip irrigation was practiced in India through indigenous methods such as bamboo pipes, perforated clay pipes and pitcher/porous cup irrigation. The bamboo made long hollow pipes of varying diameter (50-100cm) are used in making channels by the tribal farmers of Meghalaya for drip irrigation to betel, pepper and areca nut crops. The source of water is hill streams, which are diverted to hill slopes and the discharge rate at the head varies from 10-30 lit/min. and is reduced to 10-30 drops per minute at the time of application. Perforated earthenwares were popular in Maharashtra. Earthen ware pitchers and porous cups were popular in Rajasthan and Haryana for growing vegetable crops (Fig 1.1). The technique of using these is embedding of the earthen cups of about 500ml capacity at the side of the plant. The cups are filled with water at 4-5 days interval. This practice advantageously can be used for the farmers of small plots.

Figure 1.1: Earthenware pitcher irrigation

Drip irrigation was introduced in India in the early 70's at the Agricultural Universities and other research Institutes. Significant development has taken place only in 80's. The scientists of Tamil Nadu Agricultural University are the pioneers in drip irrigation research. They have conducted experiments and farm trials in the farmers' field for various crops such as vegetables, grapes, banana, cotton and sugarcane. The progressive farmers in the state of Maharashtra, Karnataka, Tamil Nadu and Andhra Pradesh have adopted drip irrigation mainly for horticultural crops. The farmers were readily convinced due to use of locally available pipes or micro-tubes and providing them the scope of cultivating more area from the available little water in the wells. The progress of drip irrigation has really gained momentum in the recent years. From a mere 1500ha in 1985 it has grown to 6000ha in 1988, 70,859ha in 1994 and 2, 59,500ha in 2000 (Table 1. 1). So far, maximum drip area covered by tree plants (42.2%) followed by vegetables (12.50%) and vines (13.20%) (Table 1.2). The total cropped area suitable for micro irrigation in the country is 27 Mha (Table 1.3). The crops found suitable for drip irrigation is listed in Table 1.4.

Table 1.1: Area covered by drip irrigation in India

State	Area (ha) in 1995	Area (ha) in 2000
Andhra Pradesh	11,585	31,600
Assam	180	200
Gujrat	3,560	8,000
Haryana	120	1,900
Karnataka	11,412	40,000
Kerala	3,035	6,000
Madhya Pradesh	1,415	3,000
Orissa	-	2,800

State	Area (ha) in 1995	Area (ha) in 2000
Punjab	-	1,500
Rajasthan	304	30,000
Tamil Nadu	5,357	34,000
Uttar Pradesh	111	2,000
West Bengal	100	200
Others	756	2,000
Total	70,859	2,59,600

Source: A.Alam & A.Kumar (2000)

Table 1.2: Percent of area covered by different crops under drip irrigation

Sl. No.	Crops	%
1	Vines	13.20
2	Vegetables	12.50
3	Field crops	7.0
4	Flowers	1.5
5	Tree crops	42.20
6	Others & unspecified	23.60

Source: A.Alam & A.Kumar (2000)

Table 1.3: Theoretical potential for drip irrigation

Crops	Area (Mha)	Area suitable for micro irrigation (Mha)
Cereals & millets	100.4	00.00
Pulses	22.50	00.00
Sugarcane	4.10	
Condiments& Spices	2.19	1.40
Fruits	3.40	3.40
Vegetables	5.30	5.30
Coconut	1.90	1.90
Oil seeds	26.20	1.90
Cotton	9.00	9.00
Others	1.40	00.00

Source: Singh, H.P. (2000)

Table 1.4: Crops grown under drip irrigation

I. Cereals	1.Corn 2. Sorghum 3. Wheat	26.Ber 27.Betelvine 28.Boysen berr	52.Valencia Orange 53.Watermelon **VIII.Plantation Crops**
II. Flowers	4.Chrysanthemum 5.Camation 6.Jasmine 7.Rose 8. (All) Ornamental Trees & Shrubs	29.Cherry 30.Chikoo(Sapota) 31.Citrus 32.Custard apple 33.Fig 34.Grape (Table & Wine)	54.Bamboo 55.Cocoa 56.Coffee 57.Mulberry 58.Olipalm 59.Rubber 60.Sugarcane
III. Fodders	9. Alfalfa 10. Asparagus 11. (All) Pastures	35. Grape fruit 36.Guava 37.Lemon	61.Tamarind 62.Tapioca
IV. Fibres	12. Cotton 13. Sisal	38.Lime 39.Mango	
V. Nuts	14. Almond 15. Arecanut 16. Cashewnut 17. Coconut 18. Macadmaia 19. Walnut	40.Mosambi 41.Naval Orange 42.Papaya 43.Peach 44.Pear 45.Pineapple	
VI.Oilseeds	20. Groundnut	46.Persimmon	
VII.Orchards	21. Amla 22. Apple 23. Aricot 24. Avocado 25. Banana	47.Plum 48.Pomegranate 49.Strawberry 50. Tangelo 51. Valencia Orange	

*Source:*INCID (1994)

In 1981, Government of India constituted a National Committee on the use of Plastics in Agriculture (NCPA) for the purpose of promoting and developing the plastics in Agriculture and Irrigation towards increasing agricultural production and irrigation efficiency. In 1983, during the National Seminar at Coimbatore it was suggested to establish Plasticulture Development Centre to collect, assemble and coordinate the research and field base information. It was further suggested that having established the technical feasibility and economic viability, the state department of agriculture and horticulture might be included for popularizing the drip irrigation through their extension programmes. Indian Petro Chemical Limited (IPCL) in collaboration with Agricultural Finance Corporation (AFC) initiated a techno-economic feasibility report on drip irrigation on pilot scheme basis in Rajasthan, Tamil Nadu, Karnataka, Gujarat, U.P. and Bihar covering about 3,000ha in about 400 villages.

The National Seminar conducted by NCPA during 1987 deliberated on the performances of the system and strategy on further promotion. Due to adequate technical and raw material support by the IPCL attracted the entrepreneurs to put up drip manufacturing units. They offered the farmers all kinds of pre-sale and after-sale services, which caused tremendous effect in increasing the drip area. The integrated approach of NCPA helped PDC's in the country and IPCL helped in sustained development of the center by involving the farmers and the other organizations such as Government Agricultural Departments, NABARD, Banks, Volunteer Organizations and Universities in the process of development.

Ministry of Water Resources (Minor Irrigation Department), Govt. of India initiated a centrally sponsored subsidy scheme in the year 1982-83 of the 6th five year plan for the purpose of encouraging the use of water saving devices like sprinklers, drip system, solar pumps, wind mills, hydrants, water turbines and man or animal operated pumps. Central Govt. provided a subsidy of 50% to the farmers with a matching contribution of 50% from state Governments for installation of the devices. Of the total subsidy, 75% was meant for small and marginal farmers with the balance of 25% for the other farmers. The above subsidy scheme continued till the 7th plan period with some modification to subsidy rates. To promote the micro-irrigation system and solar pumps, the schemes like Use of Plastics in Agriculture, Oil Palm Development Program and Integrated Central Development Programme on Sugarcane are being implemented in the country through Department of Agriculture and Cooperation, Govt. of India. In all these schemes the farmers were provided the assistance to the extent of 90% of the capital cost of the system for a hectare or Rs.25, 000/- per hectare or whichever is less for SC/ST, small/marginal and women farmers, 70% of the cost of other categories of farmers. The cost of incentive is shared 90% by the Central and 10% by State Department. In 8th five year plan the Ministry of Agriculture initiated the program for promoting the drip irrigation as a Centrally Sponsored Scheme on Use of Plastic in Agriculture. Besides, program for micro irrigation was taken up through different schemes like Technology Mission for Integrated Development of Horticulture in North East (TMNE). Integrated Scheme for Oil Seeds, Pulses & Oil Palm and Maize (ISOPOM). Despite all these efforts, the coverage of area under micro irrigation was only about 2 million hectare whereas the Task Force on Micro Irrigation (2004) indicated a potential of 69 million hectare (Anonymous, 2006). Ministry of Agriculture, Government of India modified the micro irrigation scheme in 10th five-year plan period in respect to subsidy and mode of implementation. The 40% of the scheme cost was borne by the Central Government, 10% by the State Government and the remaining 50% by the beneficiary, either through his/her own resources or soft loan from financial institution. Maximum financial assistance was limited to five hectare area per beneficiary family. The implementation of the scheme involved District Rural Development Agencies (DRDA), ICAR Institutes, SAUs, NGOs and Panchayat Raj Institutions.

Components of Drip Irrigation System

The basic essential components of a drip system consist of a pump, distribution lines (main, sub-main, and laterals) and drippers. For better control and monitoring the irrigation, the system also includes the equipments, viz. valves, pressure regulators, filters, pressure gauzes, fertilizer applicator, etc (Fig.1.2).

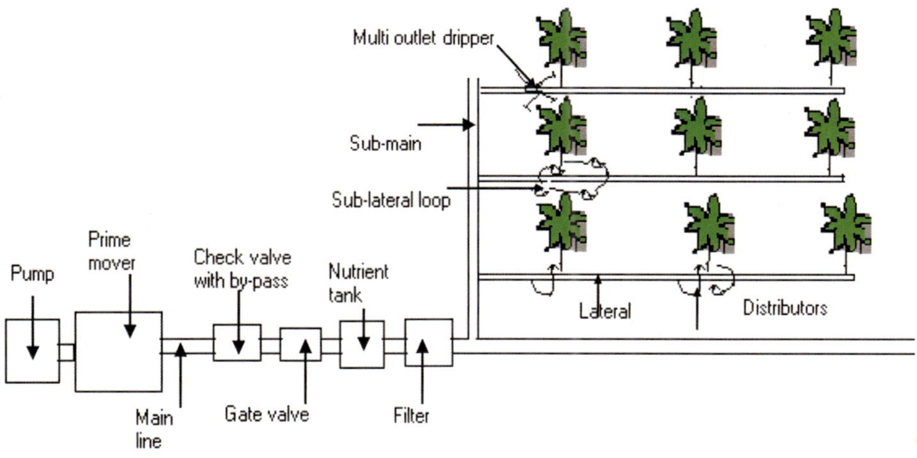

Figure 1.2: Basic components of a drip irrigation system

Pump and prime mover

A pump of suitable capacity is used to supply water through the components of the system at certain level of pressure. The source of water is usually a tank. However, groundwater can also be used directly to the drip irrigation system. If the source of supply is natural stream or farm pond there is possibility of organic and inorganic foreign bodies in water. In such case suction filter should be used for obtaining comparatively clean water. The diesel engines or electric motors are the common prime mover of the pump. In recent time the solar pump is being tried to popularize it for drip irrigation purpose. Usually the centrifugal pumps are used; however, for small system a piston pump can be used.

Control head

The control head of the drip system is responsible to regulate the pressure and water supplied, filtering of water, and addition of nutrients in it. This component includes the fertilizer applicator (tank), filter, and some control valves.

Fertilizer tank: Fertilizer tank is used to add suitable nutrient in drip water, especially nitrogen. This enables direct application of fertilizer with irrigation water and reduces the requirement of fertilizer use. The tank is a small vessel having inlet and outlet connected to the main line. As shown in Fig.1.3 a portion of the flow is directly diverted to the tank to dissolve the nutrients and further join to the main line through the outlet. The point at which the tank is connected to the main line is sometime a venturi. This increases the velocity head and develops the suction to force tank water in to the main line forming the control head of drip irrigation system

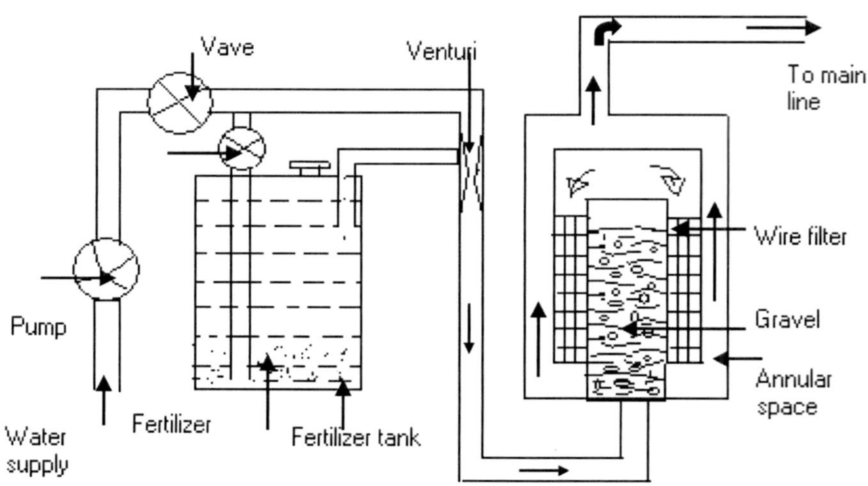

Figure 1.3: Sketch showing the arrangement of pump, fertilizer tank, filter unit and accessories

Filter: There should be a good quality filter in control head installation of a drip system. The filter uses to clean the suspended impurities of water supplied by the pump before it reaches to drippers. Impurities in irrigation water may cause blockage the holes and passage of drippers. Success of drippers is greatly depending on the performance of filter.

Distribution lines

Main line: The main line carries the total amount of water for the irrigation system. It connects the different sub-mains to water source. The main pipes are commonly made of flexible material such as PVC (poly vinyl chloride) or plastics. However, the rigid pipe of asbestos cement or galvanized steel is also used similar to main line for conventional sprinkler irrigation.

Sub-main: The sub-main feed to the laterals on one or both sides. It is made of either of medium density polyethylene (PE) or of PVC. There should be balance between the diameter of main and sub-mains. These are determined in consideration to rate of discharge, number of sub-mains, and friction losses in pipes.

Laterals: It is more commonly made of low density PE of usual diameter 1 to 1.25cm. The 1.2cm diameter laterals are popularly used. In some exception cases the small diameter rigid PVC pipe laterals are found in use. The distributors are connected to predetermine spacing in the laterals or near the plants in the case of orchards. The individual lateral length is usually limited to 40m and a pressure drop of maximum 10 percent between the two ends of a lateral.

Distributors: The distributors drip the water at low discharge rate and at atmospheric pressure. The distributors may be a dripper or a nozzle, a micro tube or any type of commercially manufactured outlet.

Types of Drip System

The following are the types of drip system.

Surface drip

Surface drip system is one where emitters and lateral lines are laid on the soil surface. This is the most common and popular type of drip system. It is suitable for wide spaced plants as well as row crops. The discharges are less than 12 lpm for single outlet point-source emitters and 12 lpm for line source emitters. The advantages of using surface drip lies in its easy installation and inspection, changing and cleaning emitters, scope of observing surface wetting pattern, and measurement of individual emitter discharge rates.

Sub-surface drip system

In this system water is slowly applied below the surface through the emitters. Sub-surface drip should not be confused to sub-irrigation in which root zone is irrigated by controlling the water table. Sub-surface drip systems have gained wider acceptance due to removal of earlier problems of clogging at large extent. It has been found that emitters pointed upward perform better. Sub-surface drip system provides little interference with cultivation or any cultural practices and possibly longer operational life. Sometime the surface and sub-surface methods both are applied partially in same field by burying the laterals and placing the emitters on the surface by the riser tubes (sub-laterals). Simply using perforated/porous pipes can also develop sub-surface drip system. Porous sub-surface irrigation is, "application of water below the soil surface at the root zone of plants through tiny openings provided on the wall of the pipe at a rate that the soil to absorb water at its natural rate". The porous pipe is made of recycled rubber and polyethylene. This pipe allows both air and water to pass through pores provided the wall at low pressure. The tiny openings in pipes are inbuilt and not mechanically made holes.

Bubbler

In bubbler irrigation water is applied in the form of small stream or fountain from a point source and the discharge rate greater than the surface and sub-surface drip system but usually less than 225 lpm. This rate of application exceeds the infiltration rate. Therefore, a basin is required around the plant to control the distribution of water. Bubbler system requires reduced filtration repair and maintenance and less energy in comparison to surface and sub-surface drip system.

Spray

In this system water is applied to the soil surface in the form of small spray, jet, fog, or mist. Air is the media of distribution of water instead of soil in case of surface, sub-surface or bubbler. The rate of application is usually less than 175 lpm and used to irrigate tree or wide spaced crops. Spray system is vulnerable to high wind and evaporation loss. Its advantages lie in minimal filtration and repair and maintenance requirements.

Mechanical move

This system is furthering of bubbler concept to large-scale row crops where water is applied through traveling drip or spray and drag or hose-reel drip system. In a traveling drip system linear-move sprinkles are used excepting the sprinkler devices but the attached lateral lines (having the linear move) use to deliver water as a continuous moving stream to each row. Uniformity of water application is good and pressure requirement is less in this system than the conventional sprinkler system. However, it requires preventive measures to check soil erosions and runoff as because the rate of application is usually more than infiltration rate. Hose-reel or drag-type drip irrigation system use to pull the surface drip system as a set across the field from one row to the next. This method is suited to use as supplemental irrigation practices.

Advantages of using mechanically moving drip irrigation system are the possible reduction in clogging problems and less expensive pipe network in comparison to surface or sub-surface drip system. The disadvantages are high initial cost, limited water application and extensive maintenance requirements.

Pulse

The pulse drip system uses high discharge rate emitters and consequently short duration application. This has the application cycle of 5, 10, or 15 minutes in an hour and the discharge rate of the pulse emitter are usually 4 to 10 times than the discharge rate of the typical emitters. Pulse drip system provides the advantage of possible reduction in clogging and disadvantage is the requirement of reliable inexpensive pulse emitter and automatic controller. Pulse drip system may also cause the startup and shutdown inefficiencies due to increase of number of application cycle.

Advantage and Disadvantages of Drip Irrigation

Advantages

Easier management

Drip irrigation does not impede the other farm operations, viz. spraying, harvesting, processing, weeding, etc. Irrigation and farm operations can be done simultaneously. Farmers have the easy access to the crop fields at all time. It is not possible in other methods of irrigation. Drip irrigation has provided easier management scope.

Saving in water

It is a general agreement that water requirement is less in drip irrigation compared to conventional method of irrigation. However, the saving of water depends on crop, soil, environment, and efficiency in farm management. The water saving may usually range 30 to 60 percent depending on the area covered by the plants and level of management of conventional irrigation (Table 1.5&1.6). The reasons behind the saving of water in drip irrigation may be listed as:

a. Less surface is wetted
b. Less water is lost due to evaporation
c. Water is distributed evenly throughout the root zone
d. No percolation losses
e. No water is utilized by weeds

Table 1.5: Yield increase and water saving under drip irrigation

Crop	Yield (t/ha)			Water use (mm)		
	Conventional	Drip	Yield increase (%)	Conventional	Drip	Water saving (%)
Banana	57.5	87.5	52	1760	970	45
Grapes	26.4	32.5	23	532	278	98
Sweet lime	100	150	50	1660	640	61
Pomegranate	55	109	98	1440	785	45
Papaya	13.4	23.5	75	228	74	68
Tomato	32.0	48.0	50	300	184	39
Watermelon	24.0	45.0	88	330	210	36
Ocra	15.3	17.7	16	54	32	40
Cabbage	19.6	20.0	2	66	27	60
Chilli	4.2	6.1	44	110	42	62
Sweet potato	4.2	5.9	39	63	25	60
Beet root	46	49	7	86	18	79
Raddish	70.0	72	2	46	11	77
Sugarcane	128	170	33	2150	940	56
Cotton	2.3	2.9	26	90	42	53

Source: Singh *et al.*. (1993)

Table 1.6: Water losses under various irrigation methods

Method of irrigation	Normal climate	Hot climate
Surface	30-45%	35-50%
Gate pipe	15-20%	20-25%
Sprinkler	6-9%	10-20%
Drip	1-2%	2-3%

Source: Saksena (1993)

Saving in labor

There is great saving in labor in drip irrigation compared to conventional irrigation. However, the system operates in low labor input only when it is properly designed, correctly installed and good quality filtered water is supplied. Saving in labor is very attractive to the countries where labor is scarce and expensive. Substantial (60-90%) saving in labor occurs as the system eliminates the need for constructing borders, bunds and other labor intensive works associated with traditional or sprinkler irrigation.

The saving of energy comes from the reduction in requirement of water to be pumped. The drip system operates at low pressure in comparison to other pressurized irrigation system. This reduces the cost of energy. However, only the efficient irrigation can claim the significant saving in energy over the conventional.

Increase in plant growth and yield

The soil moisture level at plant root zone remains fairly constant at drip irrigation because water is applied frequently and slowly at a predetermined level. There is wide fluctuation in soil moisture content in conventional and some sprinkler irrigation. Frequent drip irrigation ensures increase in yield and plant growth due to optimum soil moisture and high temperature, good aeration, better environment for root development, and reduction in disease factors. Plant growth in drip irrigation is also influenced by the better water distribution along the row and water holding capacity in heterogeneous soils.

Improved fertilizer and other chemical application

Application of fertilizer with drip water has been proved beneficial for many crop production situations. Higher fertilization efficiency is obtained as (i) water is applied frequently which match the plant requirement of fertilizer at various growth stage, and (ii) good distribution of fertilizer and merely no loss due to leaching or runoff. The fertilization/chemical study has shown a saving of 30-40% of fertilizer/chemical (Singh, 2000).

Use of saline water

Frequent application of water in drip system keeps salts in soil water more diluted and below the damaging level (Fig.1.4). This overcomes the problem of using saline water with conventional methods where great fluctuation of soil-moisture content causes to plants salt affected. Excepting this soil water can be used because of continual water feeding forces the salt concentration in the side and under the root zone, less water carries less salt to the soil, less evaporation means less salt deposit, and 20-30cm rain can flush the deposit. The study of drip and furrow irrigation with saline water on radish and potato showed much increase in yield and water use efficiency with drip irrigation (Table 1.7).

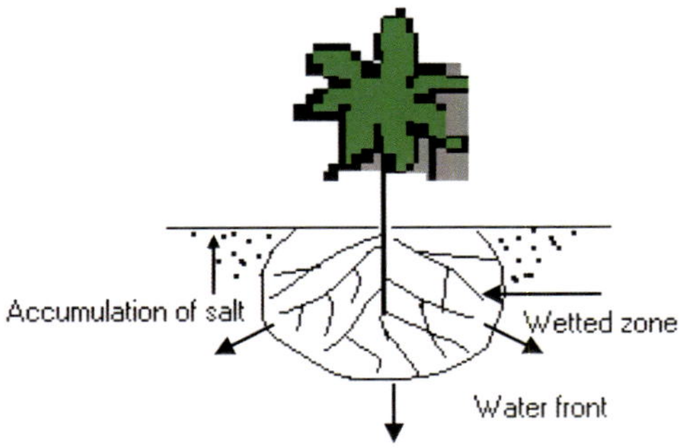

Figure 1.4: Accumulation of salt in drip system

Table 1.7: Yield (t/ha) and water-use efficiency (WUE) (kg/ha-cm) for different vegetable crops under drip and surface method of saline irrigation.

Irrigation	Reddish		Potato	
	Yield	WUE	Yield	WUE
	EC_{iw} = 6.5ds/m		EC_{iw} = 4.0ds/m	
Drip method	15.7	17.5	30.5	93.5
Furrow	9.9	8.7	19.2	53.6

Limited weed growth

Only a portion of soil surface is irrigated in drip irrigation. Therefore, weeds do not get scope to grow covering the entire area. Filtered water in drip irrigation also allows limited weed seeds to field in comparison to other methods. Investigators are convinced to overall reduction in weed control but some of them have reported the experience of increase in weed growth and control problems within the small wetted portion of the soil surface, particularly at the early days of crop.

Better use of poor soils

Light soils cannot be irrigated by surface method. It will cause high percolation loss. Heavy soils cannot be irrigated by sprinkler due to poor infiltration rate. Drip systems can be successfully used in both the situations.

Easier control of pests and weeds

Due to all time access to the field and dry foliage and soil surface favours easier, efficient and effective control of pests and weeds.

No soil erosion

Water application is much less (less than the infiltration rate) in drip irrigation. Therefore, there is little scope for soil erosion, which takes place in other conventional methods.

Flexible to operate

Spacing of the laterals, drippers, etc. can be adjusted as well as it can be shifted and accommodated in varying requirement in other fields after using it in any field for a season or so.

Disadvantages

i. Initial cost is high
ii. Technical knowledge is required to design, installation and maintenance
iii. Clogging of emitters is a great problem in many areas
iv. Assured source of water is required
v. Inadequate customer services for maintenance and supply of spare parts

Evaluation and Futuristic Approach of Drip Irrigation in India

In 1997-98 Agricultural Finance Corporation Limited evaluated the of Government subsidized scheme on drip/micro irrigation by covering 3,900 farmers of 26 districts in Andhra Pradesh, Haryana, Karnataka, Maharashtra, Orissa and Tamil Nadu. The study indicates that the beneficiary invariably introduced high-value horticultural crops like grapes, banana, mango, cashew nut, and coconut after installing drip system. The increase in yields recorded 41% for grape, 14% for pomegranate over the state (Maharashtra) average yield. Economic analysis of 695 beneficiary farmers (who installed drip system without any Govt. subsidy) indicated that the cost of investment (inclusive of subsidy was recovered within the period of less than 3 seasons in most of the cases. The development of drip irrigation is not well or even fairly distributed through the state. In Andhra Pradesh the drip irrigation was concentrated in the districts of Anthapur, Chittor, Rangareddy, Medak, Mahboob Nagar, and Nalygonda. Similarly, the coverage was largely in the districts of Chegalpattu-MGR Dindigal, Coimbatore and Periyar. In 1993 a survey was conducted by Water and Land Management Institute (WALMI) of Maharashtra over 12 randomly selected farms and found that emission uniformity was 85% or better in 4 farms but the values were 50% or less in half of the sample (Singh, 2000).

Sivanappan (2000) summarized the benefits of drip irrigation in India as below:

i. Gap between the irrigation potential is created and utilized is narrowed down.
ii. Losses due to transmission and evaporation acquiring in the open channel distribution avoided.
iii. Simultaneous irrigation is assured to all farmers.
iv. Volumetric changing in water supply could be easily done.

v. By introducing drip irrigation in canal command, crop diversified system is introduced.

vi. The system provides equitable distribution of water from the head to tail-end farmers.

vii. Over-exploitation of ground water can be controlled.

viii. Less electricity power requirement.

ix. Creates avenues for establishing agro-based industries.

x. Overall agricultural production is increased, improving socio-economic status of farmers.

xi. Fertilizer application will be minimized through fertigation.

In spite of very good result and attractive subsidy the adoption of drip irrigation still lagging much behind to expected level. Sharma & Alam (1993) found the apparent reasons as listed below:

i. High initial cost of the system.

ii. Necessary of an assured irrigation water sources.

iii. These high-tech irrigation methods are energy intensive.

iv. Poor quality of material and maintenance services.

v. Limited choice and marketing facilities for the high value crops.

vi. Poor technical knowledge of the agricultural extension agencies and absence of irrigation information services.

vii. Attractive only in areas where water availability is low/cost of water is high.

viii. Lesser emphasis trials on identification of target areas and promising irrigators.

ix. Few adaptive trials and operational research projects in different agro-ecological regions.

x. Lack of research efforts on developing cost-effective designs suited to local conditions.

xi. Present water pricing acts as a deterrent to adoption of these systems in irrigated areas.

There are many scientists who worked on to identify the steps to be undertaken to popularize the drip irrigation in India. Samuel & Singh (1998) emphasized on the following:

i. Supply of standard materials and equipment confirming to BIS standard.

ii. The prompt customer services for maintenance and supply of adequate spares.

iii. The annual coverage of area under drip irrigation in the country needs to be enhanced to at least 50, 000ha against the current level of 30,000ha.

iv. Efforts would be needed to ensure the balanced development of drip irrigation in all the parts of the country.

v. The state implementing agencies also need to gear up the tempo, particularly in Gujarat, M.P., Rajasthan, Punjab and U.P.

vi. Studies are needed in hardware aspects of drip installation including the development for indigenous emitters. Use of micro sprinklers in an area for which data need to be generated.

vii. The program for training the field functionaries need to be strengthened.

viii. Database of area brought under micro irrigation under different crops need to be developed.

ix. There is need to popularize fertigation to economize the use of fertilizers.

Rande (2000) emphasized on the followings:

i. Educating policy makers, irrigation engineers, agricultural engineers etc. through workshops and seminars on micro irrigation system.

ii. For the irrigators/cultivators, demonstration farms and visit to micro-irrigation, high- value crops farms should be visited.

iii. For the manufacturers, standardization and certification of quality of product, providing excellent after sales service should be made statutory obligatory.

iv. Encouraging adaptive research in micro irrigation and publication of realistic and actual field studies of economic viability of micro irrigation system is essential. Publication of brochures furnishing typical designs of the system useful to common cultivators would be helpful.

v. At the Govt. level, policy decisions should be taken about giving concession/ tax benefits to hardware manufacturers, setting up of quality control standards and laboratories for testing of equipments, ensuring efficient after sale service, provide just adequate subsidies to irrigators, providing better marketing facilities, export possibilities to additional produce, etc.

1.0 Questions & Problems

1.1. Define drip irrigation. Give some other terms of drip irrigation.

1.2. Discuss the history of drip irrigation in India and abroad.

1.3. Explain the drip irrigation system with necessary diagram.

1.4. Explain different type of drip system.

1.5. What are the advantages and disadvantages of drip system?

1.6. What are the reasons behind the saving of water in drip irrigation?

1.7. Discuss the state of drip irrigation in India.

1.8. Describe the futuristic approach of popularizing drip irrigation in India.

1.9. What are the major components in a drip system? Describe in detail.

1.10. Discuss the drip irrigation with reference to easier management and better use of poor soil and water.

1.11. Discuss the reasons behind not satisfactory progress of drip irrigation in India.

1.12. Subsidy to drip irrigation is in fact very good investment- explain.

1.13. Write True or False of the following statements

1. Drip irrigation encourages evaporation loss.
2. Drip irrigation is essentially frequent and long duration application of water.
3. Pitcher irrigation may be said as indigenous method of drip irrigation.
4. Distributors drip the water at higher rate of atmospheric pressure.
5. The diameters of lateral pipes are usually 10 to 15 cm.
6. Spray, fog, mist are the forms of drip irrigation.
7. Drip irrigation is more attractive to the area where cost of water is high.
8. Sub-surface drip system provides little interference with cultivation.
9. In drip irrigation mechanical move is the furthering of bubbler concept.
10. Fertilizer applicator is not a component of control head in drip system.

Ans. 1. False 2. True 3. True 4. False 5. False 6. True 7. True 8. True 9. True 10. False

1.14. Select the appropriate answer from the following.

1. Bubbler irrigation apply water in the form of

   a) Trickle b) Capillary rise

   c) Small stream d) Jet fog

2. Modern drip irrigation was introduced by

   a) Samuel Jose b) Keler & Karmelli

   c) Freeman & Gazzoli d) Symcha Blass

3. Emitter is said to be
    a) Distributor   b) Lateral
    c) Pressure regulator   d) Fertilizer tank inlet
4. ASAE used to state micro-irrigation as
    a) Trickle irrigation   b) Drop irrigation
    c) Drip irrigation   d) Localized irrigation
5. Drip irrigation was introduced in India
    a) 60's   b) Early 60's
    c) Early 70's   d) Late 70's
6. Drip irrigation save water to the range of
    a) 10-20%   b) 10-30%
    c) 20-50%   d) 30-60%
7. Pulse drip system provides the advantage of reduced
    a) Evaporation   b) Clogging
    c) Weeding   d) Percolation
8. Frequent drip irrigation ensures
    a) Saturated soil moisture   b) Optimum soil moisture
    c) Low soil temperature   d) Wider water spreading
9. Heavy soil cannot be irrigated by sprinkler due to
    a) Hhigh water holding capacity
    b) High saturation percentage
    c) Poor infiltration
    d) High porosity
10. Porous sub-surface irrigation applies water below the soil through
    a) Buried small diameter laterals
    b) Buried efficient distributors
    c) Sub-surface spray
    d) Buried pipes with tiny openings on the wall

Ans. 1. (c) 2. (d) 3. (a) 4. (c) 5. (c) 6. (d) 7. (b) 8. (b) 9. (c) 10. (d)

References

Anwar Alam & Aswani Kumar (2000). Micro-irrigation system-past, present and future. Proceedings of International Conference on Micro and Sprinkler Irrigation Systems, 8-10 Feb., 2000, Jain Irrigation Hills, Jalgaon, Maharashtra, India. pp.1-17.

INCID (1994). Drip Irrigation in India. pp.115.

Ranade. R.S. (1993). Flow irrigation to micro irrigation, a journey on path of prosperity and equity through optimum use of scarce water resource. Proceedings of Workshop on Sprinkler and Drip Irrigation Systems, held during 8-10 December, 1993 at Jalgaon, India. pp.38-43.

Saksena, R,S. (1993). Sprinkler and drip irrigation in India- present bottleness and suggested measures for speeder development Proceedings of Workshop on Sprinkler and Drip Irrigation Systems, held during 8-10 December 1993 at Jalgaon, India. pp.26-40.

Samuel, Jose C. and Singh, H.P. (1998). Current trends of micro-irrigation development in Indian horticulture. Proceedings of Workshop on Micro Irrigation and Sprinkler Irrigation System, held at CBIP, New Delhi, during 28-30 April, 1998. pp.1-13.

Singh, H.P. (2000). Emerging scenario of micro irrigation in India. Proceedings of International Conference on Micro and Sprinkler Irrigation Systems, 8-10 Feb., 2000, Jain Irrigation Hills, Jalgaon, Maharashtra, India. pp.18-30.

Singh, J., Singh, A.K. and Garg, R (1995). Present status of drip irrigation in India. Proceedings of Workshop on Sprinkler and Drip irrigation Systems, held during 8-10 December, 1993 at Jalgaon, India. pp.11-15.

Sharma, B.R. and I.P.Abrol (1993). Proceedings of Workshop on Sprinkler and Drip Irrigation Systems, held during 8-10 December, 1993 at Jalgaon, India. pp.21-23.

Sivanappan, R.K (2000). Scope and need for introducing micro irrigation in Parambikulam and Aliyan Project (PAP) in Tamil Nadu. Proceedings of International Conference on Micro and Sprinkler Irrigation Systems, 8-10 Feb., 2000, Jain Irrigation Hills, Jalgaon, Maharashtra, India. pp.63-68.

2

Hydraulics of Flow Regime

In drip irrigation water is distributed throughout the field from a system of pressurized pipelines. The pressure must be sufficient to overcome the frictional losses and elevation differences. The system also deserves discharge from all the emitters and the emitters must dissipate the pressure difference between the inlet to emitters and atmospheric pressure.

The flow through the drip pipes may be laminar or turbulent or a combination of both. The laminar flow is characterized by the layers of cylindrical tubes of which the maximum velocity is at the center about double of the average velocity. The turbulent flows having *pulsatory* cross current velocities and maximum velocity is about only 1.25 time of the average velocity. Turbulent flows loss more energy than laminar flow due to development of the cross current velocity.

Reynolds Number

The criterion for distinguishing the flow from laminar to turbulent was developed by Osborne Reynolds (1842-1900) and was named as Reynolds number; R_e. The Reynolds number in circular pipes flowing full is expressed as

$$R_e = \frac{vd}{v} \qquad (2.1)$$

Where, v = the average velocity of flow, m/s

d = the diameter of pipe, m

ν = kinematics viscosity of water, m^2/s

The pipe diameters in drip system are usually low and expressed in millimeters. Therefore, the Eq.5.1 may be rewritten by

$$R_e = \frac{vd}{Kv} \qquad (2.2)$$

Where, the diameter of the pipe is in millimeter and K is a constant, which equals to 1000.

The Eq.2.1 may be written as below by introducing the discharge, q

$$Re = \frac{(q/a)d}{v}$$

$$= \frac{qd}{\frac{\pi d2v}{4}}$$

$$= \frac{4q}{\pi dv} \tag{2.3}$$

If the discharge (q) is taken in l/h and diameter (d) in millimeter, the Eq.2.3 becomes

$$Re = \frac{4q/3.6x10^6}{1000\pi dv}$$

$$= \frac{4q}{K\pi dv} \tag{2.4}$$

The value of constant K is equal to 3600.

Darcy-Weisbach Equation

The loss of energy in pipes can be calculated by the well-known Darcy-Weisbach equation as

$$h_l = f\frac{l}{d}\frac{v^2}{2g} \tag{2.5}$$

Where, h_l = the energy loss in a pipe of length l

f = the friction coefficient

d = the internal diameter of pipe

v = the average velocity

g = acceleration due to gravity.

By introducing the discharge q, Eq.2.5 is become

$$hl = f\frac{l}{d}\frac{q^2}{\left(\frac{\pi d^2}{4}\right)^2 2g}$$

$$= 8f\frac{l}{d^5}\frac{q^2}{\pi^2 g} \tag{2.6}$$

The friction coefficient f depends on R_e and the relative roughness of pipe wall (Table 2.1).

Table 2.1: The flow regimes as defined as a function of R_e and the friction factor f

Flow regime	**Reynolds number, R_e**	**f**
Laminar	$R_e \leq 2{,}000$	$f = \frac{64}{R_e}$
Unstable	$2{,}000 \langle R_e \leq 4{,}000$	$f - 3.42x10^{-5}\, R_e^{0.85}$
Partially turbulent	$4{,}000 \langle R_e \leq 10{,}000$	$f = \frac{0.3164}{R_e^{0.25}}$
Fully turbulent	Re>10,000	$f = \frac{0.3164}{R_e^{0.25}}$

Darcey-Weisbach equation for different flow regimes

Laminar flow

If flow is laminar, $f = \frac{64}{R_e}$ (2.7)

The friction coefficient f is independent to relative roughness of pipe. It only depends on R_e. By introducing Eq.2.7 in Eq.2.6

$$h_l = \frac{64}{R_e}\frac{l}{d}\frac{v^2}{2g}$$

$$= \frac{64}{vd}v\frac{l}{d}\frac{v^2}{2g}, \text{putting } R_e = \frac{vd}{v}$$

$$= \frac{32vl}{d^2}\frac{v}{g}$$

$$= \frac{32vl}{d^2}\frac{q}{g\pi\frac{d^2}{4}}$$

$$= \frac{128vlq}{\pi d^4 g} \tag{2.8}$$

Or, , $q = \frac{h_l \pi d^4 g}{128vl} = Ch_l, C = \frac{\pi d^4 g}{128vl}$ (2.9)

In Eq.2.9, C is a constant in a pipe at a given operating condition and temperature. The kinematics viscosity is greatly influenced by the change of temperature. This is known as temperature sensitivity.

Assuming the normalized flow at 20^{0}C, the flow at other temperature is multiplied by a factor. The kinematics viscosity of different temperature (after Daugherby and Pranzini, 1979) and the multiplication factors are given as below.

Temperature (^{0}C)	Kinematics viscosity (n), $10^6 m^2/sec$
0	1.785
10	1.306
20	1.003
30	0.800
40	0.658
50	0.553
60	0.474
70	0.413

Temperature (^{0}C)	Temperature factor
5	0.63
10	0.75
15	0.87
20	1.00
25	1.13
30	1.43

Courtesy: Nakayama & Bucks (1980) & Vermeiren & Jobling (1980)

Unstable flow

The Reynolds number in the range of 2000-4000 the flow changes from laminar to turbulent. The values of f are somewhat uncertain in this range. However, it may be taken as below for first approximation as shown in Table 2.1.

$$f = 3.42 \times 10^{-5} \times R_e^{0.85} \tag{2.10}$$

Turbulent flow

When the flow occurs at Reynolds number over 4,000, the value of f depend on the roughness of pipe as well as the viscosity and the density of water. The turbulent flow may be divided as below:

a. flow in smooth pipe
b. flow in relatively rough pipe
c. flow in the zone between above two

a) Flow in smooth pipe

In smooth pipe the relative roughness is unimportant and the value of f can be obtained by Blasius and Karman - Prandtl equations.

$$\text{when } R_e \langle 80{,}000, f = \frac{0.3164}{R_e^{0.25}} \text{(Blasius)} \tag{2.11}$$

$$\text{For all values of } R_e, \frac{1}{\sqrt{f}} = 2\log(R_e\sqrt{f}) \tag{2.12}$$

Biswas *et al.* (2005) studied the head losses in10mm lateral pipe and 4mm sub-lateral pipe at different pressure and depending on the Reynolds numbers, the friction factors (f) were found out by using the Eq.2.11 or 2.12. The friction losses and the friction factors were related to the following equations:

For 10mm pipe:

$$h_l = 0.0427H - 0.0003 \tag{2.13}$$

$$f = -0.0008H + 0.03 \tag{2.14}$$

For 4mm pipe:

$$hl = 0.0442H - 0.0035 \tag{2.15}$$

$$f = -0.0009H + 0.0042 \tag{2.16}$$

Where, h_l = friction loss per meter length of the pipe, m/m

f = friction factor

H = pressure head, m.

b) Flow in relatively rough pipe

The Reynolds number R_e is unimportant when turbulent flow occurs on a rough surface. The values of f may be calculated by the following equation:

$$\frac{1}{\sqrt{f}} = 2\log\frac{d}{s} + 1.14 \text{(Nikuradse)} \tag{2.17}$$

Where, s = dimension of the roughness

Since the above equation is independent to R_e value, by putting the f value in Darcy-Weisbach equation (Eq.2.6) it can be seen that the discharge q is proportional to square root of h_l and all other values are constant for a given pipe. Therefore, $q = C\sqrt{h_l}$ or $q = CH^{1/2}$ where C is a constant and H is the head loss.

c) Flow in the transition zone between smooth and rough boundaries

The value of f in this particular case is influenced by R_e as well as relative roughness of pipe. The following equation can be used for this condition,

$$\frac{1}{\sqrt{f}} = 1.14 - 2\log\left[\frac{s}{d} + \frac{9.35}{h_l\sqrt{f}}\right] \text{ (Colebrook-White)} \quad (2.18)$$

Example 2.1 Determine the characteristics of flow regime in a drip pipe at a discharge rate of 600 l/h, diameter of pipe 15mm and kinematic viscosity of water 5.6 × 10^{-7} m^2/s.

Solution: Reynolds number, $R_e = \frac{4q}{K\nu\pi d}$

$$= \frac{4x600}{3600 \text{ x } 5.6x10^{-7} \text{ x}\pi \text{ x } 15}$$

= 25,262.68, which is more than 4,000.

So, the flow is turbulent.

Example 2.2 What is value of friction coefficient f in **Example 2.1** when flow takes place through smooth pipe and relatively rough pipe of roughness dimension 0.5mm?

Solution: In Blasius equation for flow through smooth pipe,

$$f = \frac{0.3164}{R_e^{0.25}}$$

$$= \frac{0.3164}{(25264.68)^{0.25}} = \frac{0.3164}{12.61} = 0.025$$

For relatively rough pipe, using Eq.2.12 and f = 0.025

$$\frac{1}{\sqrt{f}} = 2\log\frac{d}{s} + 1.14$$

$$\text{or, } \frac{1}{\sqrt{f}} = 2\log\frac{15}{0.5} + 1.14$$

$$= 2.95 + 1.14 = 4.09$$

$$\therefore f = \left(\frac{1}{4.09}\right)^2 = 0.06$$

Example 2.3 Compare the friction coefficients following Blasius (Eq.2.11) and Biswas *et al.* (Eq.2.14 & 2.16) equations for the flow of 500 l/h and 200 l/h in 10 and 4mm pipe respectively. The inlet pressure for both the pipes is 8m. Assume kinematic viscosity of water 5.6 × 10^{-7}.

Solution

10mm pipe

$$R_e = \frac{4q}{K\pi dv}$$

$$= \frac{4 \times 500}{3600\pi \times 5.6 \times 10^{-7} \times 10}$$

$$= \frac{2000}{0.063} = 31578.36$$

Blasius equation,

$$f = \frac{0.3164}{R_e^{0.25}}$$

$$= \frac{0.3164}{31578.36^{0.25}}$$

$$= \frac{0.3164}{13.33} = 0.0236$$

Biswas *et al.*, $f = -0.0008H + 0.03$

$= -0.0008 \times 8 + 0.03$

$= -0.0064 + 0.03 = 0.0236$

4mm pipe:

$$R_e = \frac{4q}{K\pi dv}$$

$$= \frac{4 \times 200}{3600\pi \times 5.6 X 10^{-7} \times 4}$$

$$= \frac{800}{0.025} = 31578.36$$

Blasius equation, $f = \frac{0.3164}{R_e^{0.25}}$

$$= \frac{0.3164}{(31578.36)0.^{025}} = 0.0237$$

Biswas *et al.*, $f = -0.0009H + 0.042$

$= -0.0008 \times 8 + 0.042$

$= 0.0348$

Example 2.4 Determine the energy loss in a drip laminar pipe flow where Reynolds number is 1600, length of pipe 15m, diameter of pipe 16mm and velocity of water 45cm/s.

Solution: The flow is laminar, so,

$$f = \frac{64}{R_e}$$

$$= \frac{64}{1600} = 0.04$$

$$\text{The energy loss, } h_l = f\frac{l}{d}\frac{v^2}{2g}$$

Hazen-William Formula

There are some empirical formulae to pipe flow, which have been developed by using laboratory or field data though Darcy-Weisbach equation gives rational solution of pipe flow. Among these Hazen-William formula is commonly used:

$$v = 1.32CR_h^{0.65}S_f^{0.54} \tag{2.19}$$

Where, v = the velocity

C = the discharge constant

R_h = the hydraulic radius

S_f = the slope of the energy line i.e., the head loss divided by the pipe length

The values of C are given in Table 2.2.

Table 2.2: Values of C for Hazen-William formula

Description of Pipe	C
Polyvinyl chloride pipe	150
Extremely smooth and straight	140
Very smooth	130
Smooth wood and wood stave	120
Vitrified	110
Old riveted steel	100
Old cast iron	95
Old pipe in bad condition.	60 to 80
Small pipes badly tuberculated	40 to 60

Hydraulic Characteristics of Distributors

The distributors are the accessories fitted to the pipes, or the holes in the pipe for the purpose of allowing water to drip, flow out, or spray at a low and constant discharge. Hydraulic characteristics include the operating pressure, the range of operating pressure and the normal flow rates at 20^{0}C water temperature and pressure head of 10m of water excepting if any different pressure head is mentioned. A distributor should posses the characteristics of uniform and low discharge, sufficient aperture to prevent clogging and foreign matter and chemical deposits, low cost, robustness and homogeneity. Distributors may be grouped as (i) orifice or nozzle type, and (ii) long path type. Accordingly energy dissipation in the distributors follows either the long flow paths or the nozzle orifices. Whatever may be the method of energy dissipation; the discharge through a distributor can be expressed by the equation

$$q = K_d H^x \tag{2.20}$$

Where, q = the discharge of the distributor

K_d = a coefficient specific to each distributor

H = the pressure at which the distributor operates

x = an exponent, the value of which depends on flow regime.

The most common method of determining the values of K_d and x by the linear regression on the logarithm of flow and operating pressure using Eq.2.20

$$\ln(q) = \ln K_d + x \ln H$$

which is the linear form of

$$y = mz + c$$

where, y = ln(q), m = x, z = ln(H), c = ln(K_d).

A linear regression of the lnH on lnq will give the values of m and c. Since c = lnK_d then $K_d = e^c$, Eq.2.20 may be expressed as,

$$q = e^c H^m \tag{2.21}$$

For a set of discharges and pressures, the regression equation gives,

$$\ln K_d = \frac{\sum \ln q \sum (\ln H)^2 - \sum (\ln q . \ln H) \sum \ln H}{n \sum (\ln H)^2 - (\sum \ln H)^2} \tag{2.22}$$

$$x = \frac{n \sum \ln q . \ln H - \sum \ln q . \sum \ln H}{n \sum (\ln H)^2 - (\sum \ln H)^2} \tag{2.23}$$

When n=2

$$x = \frac{\ln (q_1 / q_2)}{\ln (H_1 / H_2)} \tag{2.24}$$

The emission exponent characterizes the flow of distribution.

For x=0.5, the flow is fully turbulent

x=1.0, the flow is laminar

$0.5 < x < 1.0$, the flow at intermediate stage of laminar and turbulent.

Studying the value of x corresponding to the flow regime it may be stated that the discharge variation is almost linear in laminar flow ($q=K_dH^x$, $x=1$). For a pressure variation of 10 percent the discharge variation will be 10 percent. In turbulent flow ($x=0.5$) with a pressure variation of 10 percent the expected discharge variation is only 5 percent. Therefore, to ensure the low and uniform discharge against the undulated topography and friction losses in pipes, high-pressure dissipation in distributors is required. Adopting small flow path does higher dissipation of energy in distributors. However, narrow the path higher is the risk of clogging and wider the path less the dissipation of pressure and higher flow rate. Fig 2.1 illustrates the discharge variation resulting from pressure head variation for different emitter.

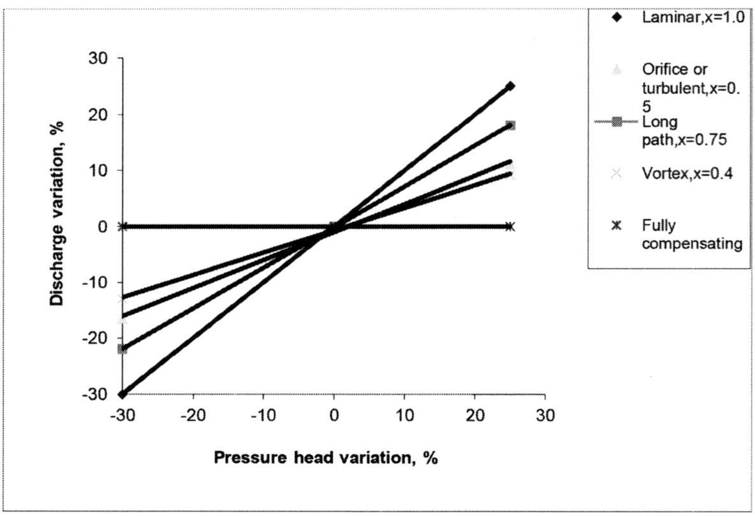

Figure 2.1: Discharge variation resulting from pressure head variation for different emitters (adapted from: Keller and Bliesner, 1990)

Flow in orifice or nozzle type distributors

Flow through any orifice or nozzles are turbulent, therefore, the value of x is 0.5 in the Eq.2.21.

$$q=K_dH^{0.5} \tag{2.25}$$

$$\text{or, } K_d=aC_d(2g)^{0.5} \tag{2.26}$$

Where, a=cross-sectional area of the orifice

C_d=a discharge coefficient depends on the characteristics of orifice or nozzle

g=acceleration due to gravity.

Flow in long path distributors

According to Darcy-Weisbach formula (Eqs.2.6 and 2.20),

$$q = K_d H^x \sqrt{\frac{Hd^5 g\pi^2}{8fl}} \tag{2.27}$$

In case of laminar flow, the equation (3.22) becomes

$$q = K_d H = \frac{\pi d^4 gH}{128\nu l} \tag{2.28}$$

The Eq.2.27 and 2.28 are expressed for turbulent and laminar flow respectively.

Example 2.5 The discharges against the pressure head recorded to a distributor are given below:

q(l/h)	0.8	1.1	1.7	2.1	2.9	3.7	4.2	4.6
H(m)	1.5	2	4	6	8	10	12	15

Develop the head-discharge relationship.

Solution

H	q	lnH	lnq	LnH.lnq	$(\ln H)^2$	$(\ln H)^2 \ln q$
1.5	0.8	0.405	-0.223	-0.090	0.164	-0.037
2	1.1	0.693	0.095	0.066	0.480	0.046
4	1.7	1.386	0.531	0.736	1.921	1.021
6	2.1	1.792	0.742	1.330	3.210	2.3832
8	2.9	2.079	1.065	2.214	4.322	4.605
10	3.7	2.3025	1.308	3.011	5.302	7.039
12	4.2	2.485	1.435	3.566	6.175	6.774
15	4.6	2.708	1.526	4.132	7.334	11.192
		∑=13.8505	∑=6.479	∑=14.966	∑=28.91	∑=33.022

$$\text{In}\,Kd = \frac{\sum \text{In}\,q \sum (\text{In}\,H)2 - \sum (\text{In}\,q\,\text{In}\,H) \sum \text{In}\,H}{n\sum (\text{In}\,H)^2 - (\sum \text{In}\,H)^2}$$

$$= \frac{6.47\text{x}28.91 - 14.966\text{x}13.8505}{8\text{x}28.91 - (13.8505)^2}$$

$$= \frac{187.3079 - 207.2866}{231.28 - 191.8364} = \frac{-19.9787}{39.4436} = -0.5065$$

$$\therefore K_d = 0.6026 \cong 0.6$$

$$x = \frac{n\sum \text{In}\,q.\text{In}\,H - \sum \text{In}\,q.\sum \text{In}\,H}{n\sum (\text{In}\,H)^2 - (\sum \text{In}\,H)^2} = \frac{8x14.966 - 6.479x13.8505}{8x28.91 - (13.8505)^2}$$

$$= \frac{29.99}{39.4436}$$

$$= 0.761$$

$$\therefore q = 0.6H^{0.76}$$

Manufacturing Variation of Distributors

The parameter that describes the anticipated variation of discharges of a set of new distributors caused by the variation in the manufacturing of the distributors is called manufacturer's 'coefficient of variation'. The manufacturing variation may occur in different way such as, inability to hold dimensional tolerances due to molding pressure and temperature, variation to material used, mold parting line flash, welding and gluing flash and mold wear. The extent of controlling the variation not only depends on the quality of the manufacturing materials and quality control but also on the mode of operation of the distributors. The most important of a distributor is to maintain the diameter of it. The diameter usually varies between 1 to 2mm, which must be manufactured precisely. With laminar flow, the flow varies with d^2 and in turbulent flow with $d^{19/7}$. Therefore, a 2% variation in d causes 4% variation in laminar flow and more than 6% variation in turbulent flow. Thus, manufacturing coefficient of flow is a great concern of turbulent flow. Reasonable ranges of CV are described in Table 2.3.

Table 2.3: Interpretation of manufacturing coefficient of variation

Coefficient of Variation (CV)	**Interpretation**
0.050 or less	Good
0.05-0.10	Average
0.10-0.15	Marginal
0.15 or more	Unacceptable

The manufacturers should provide the value of coefficient of variation. However, it can be determined from the discharge data of a set of at least 50 emitters at a reference pressure head. This is calculated as

$$CV = \frac{\sqrt{(q_1^2 + q_2^2 + \ldots\ldots + q_n^2 - n\bar{q}^2)/(n-1)}}{\bar{q}} \tag{2.29}$$

$$= \frac{\sigma}{\bar{q}} \tag{2.30}$$

Where, CV = coefficient of manufacturing variation for the set of emitters in which $q_1, q_2, \ldots\ldots\ldots, q_n$ are the individual discharge rates, l/h.

n = number of emitters in the sample

$\vec{q}$ = average discharge rate of the sample ($q_1 + q_2 + \ldots\ldots\ldots + q_n$)/n, l/h

σ = standard deviation of the discharge rate of the emitters, l/s.

The CV value has very useful physical significance, because the discharges of the emitters are normally distributed. The physical significance of a CV derived from the classic bell-shaped normal distribution curve is:

- Essentially all the observed rates fall within (1±3 coefficient of variation) (average discharge) = (1±3CV)$\vec{q}$.
- Approximately 95% of the discharge rates fall within (1±2CV)$\vec{q}$.
- The average of low one fourth of the discharge rate is approximately equal to (1-1.27CV) $\vec{q}$
- Approximately 68% of the discharge rates fall within (1±CV)$\vec{q}$.

From the above, for the emitter of CV=4% i.e., 0.04 and $\vec{q}$ =4 l/h; the maximum discharge, q_{max}= (1+3x0.04) 4 =4.48 l/h. Similarly, the minimum discharge, q_{min}=3.52 l/h. Further, 95% Of the discharge rates fall within the range of 3.68 and 4.32 l/h. The average discharge of low one fourth of the discharge is approximately 3.80 l/h. Differences in discharges due to variation in manufacturing is sometime more important than pressure variation. Therefore, it is very important to know the value of CV in selecting the distributors.

CV of multiple distributors

Distributors are used in number in trees or vines. The value of CV in such cases gets adjusted. Therefore, the variation of discharges is less than the variation to an individual distributor. The CV of multiple uses of distributors is determined as follows:

$$CV_{total} = \frac{CV}{\sqrt{e}} \tag{2.31}$$

Where, CV_{total} = the distributors' system coefficient of variation

e=the number of distributor per plant

CV=CV_{total}, when there is one outlet of the distributor.

CV & head-discharge relationship

A good number of drippers of Indian make of nominal discharges 2, 3 and 5 l/h were tested for discharge (q), average discharge $(\overline{q})$, standard deviation (σ) and coefficient of variation (CV) and is presented in Table 2.4 to 2.6. The average CV of the discharges for different pressure was 0.036, 0.026 and 0.024 for 2, 3 and 4 l/h drippers respectively. The head-discharge relationship developed for different drippers are presented in Fig. 2.2 to 2.4 and in Eq.2.32 to 2.34.

For 2l/h nominal discharge, $q = 1.3315H^{0.3092}$ (2.32)

For 3 l/h nominal discharge, $q = 1.9377H^{0.3026}$ (2.33)

For 5 l/h nominal discharge, $q = 3.2589H^{0.3409}$ (2.34)

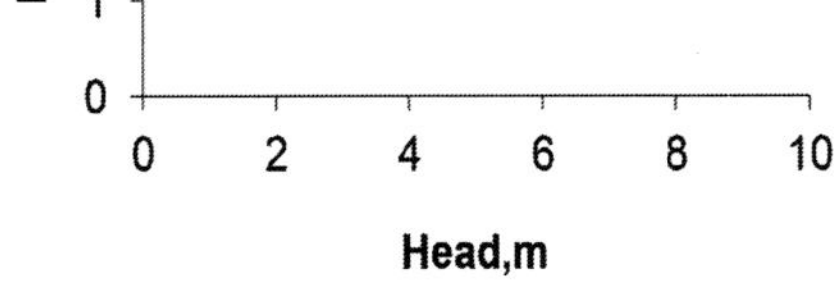

Figure 2.2: Head discharge relationship for the drippers of 2 l/h nominal discharge

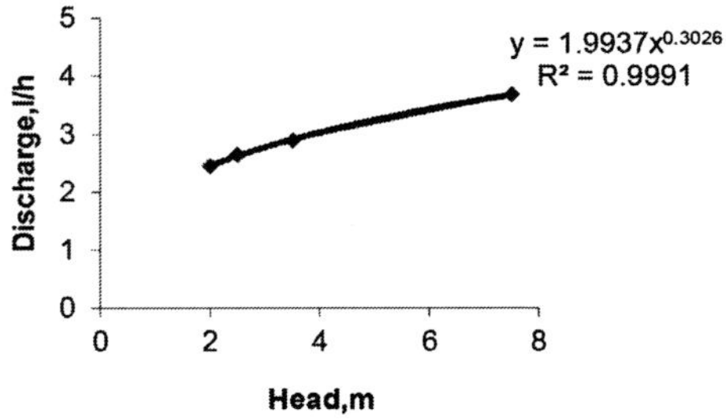

Figure 2.3: Head discharge relationship for the drippers of 3 l/h nominal discharge

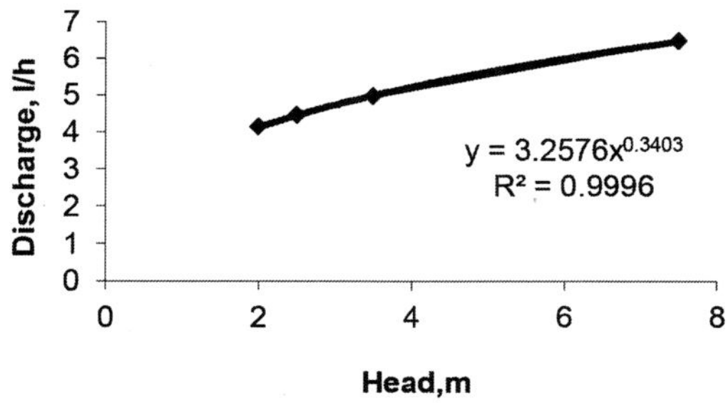

Figure 2.4: Head discharge relationship for the drippers of 5 l/h nominal discharge

Table 2.4: Discharges, CV and SD of the drippers under different pressure head at nominal discharge of 2l/h

Sl. No.	Discharges at 2m pressure head, l/h	Discharges at 2.5m pressure head, l/h	Discharges at 3.5m pressure head, l/h	Discharges at 7.5m pressure head, l/h
1	1.552	1.728	1.888	2.256
2	1.552	1.648	1.936	2.2848
3	1.552	1.696	1.936	2.304
4	1.552	1.696	1.936	2.3232
5	1.552	1.696	1.936	2.352
6	1.6	1.744	1.936	2.352
7	1.6	1.744	1.936	2.3808
8	1.6	1.744	1.9552	2.4
9	1.6	1.744	1.9552	2.4
10	1.728	1.744	1.9552	2.4
11	1.728	1.744	1.9552	2.4
12	1.728	1.792	1.9744	2.4
13	1.728	1.792	1.9744	2.4
14	1.728	1.792	1.984	2.4
15	1.568	1.792	1.984	2.4
16	1.568	1.792	1.984	2.4
17	1.568	1.792	1.984	2.4192
18	1.568	1.792	1.984	2.4288
19	1.568	1.792	1.984	2.4288
20	1.568	1.792	1.984	2.4288
21	1.568	1.792	1.984	2.4288
22	1.488	1.8208	1.984	2.4384
23	1.488	1.8208	1.984	2.4384
24	1.536	1.8208	1.984	2.4384

25	1.536	1.8208	1.984	2.448
26	1.536	1.8208	1.984	2.448
27	1.536	1.8208	1.984	2.448
28	1.536	1.8208	1.984	2.448
29	1.72	1.84	1.984	2.448
30	1.72	1.84	1.984	2.448
31	1.72	1.84	1.984	2.448
32	1.72	1.84	1.984	2.448
33	1.72	1.84	1.984	2.448
34	1.72	1.84	1.6	2.448
35	1.536	1.84	1.9936	2.448
36	1.536	1.84	2	2.448
37	1.536	1.84	2	2.448
38	1.536	1.84	2.0032	2.4576
39	1.536	1.84	2.0128	2.4576
40	1.584	1.84	2.0128	2.4768
41	1.584	1.84	2.0128	2.4768
42	1.584	1.84	2.0128	2.4768
43	1.584	1.84	2.0128	2.4768
44	1.584	1.84	2.0128	2.4768
45	1.584	1.84	2.0128	2.4768
46	1.584	1.84	2.0128	2.4768
47	1.584	1.8496	2.0128	2.4768
48	1.584	1.8496	2.0128	2.496
49	1.592	1.8592	2.0128	2.496
50	1.592	1.8592	2.0128	2.496
51	1.592	1.8592	2.0128	2.496
52	1.592	1.8592	2.0128	2.496
53	1.592	1.8592	2.032	2.496
54	1.592	1.8592	2.032	2.496
55	1.592	1.8592	2.032	2.496
56	1.552	1.8688	2.032	2.496
57	1.552	1.8784	2.032	2.496
58	1.552	1.8784	2.032	2.496
59	1.552	1.888	2.032	2.5152
60	1.552	1.888	2.032	2.5152
61	1.552	1.888	2.032	2.5152
62	1.552	1.888	2.032	2.5152

63	1.6	1.888	2.032	2.544
64	1.6	1.888	2.032	2.544
65	1.6	1.888	2.0512	2.544
66	1.6	1.888	2.0512	2.544
67	1.6	1.888	2.08	2.544
68	1.6	1.888	2.08	2.544
69	1.776	1.888	2.08	2.544
70	1.776	1.888	2.08	2.5728
71	1.776	1.8952	2.08	2.5728
72	1.776	1.936	2.08	2.592
73	1.728	1.936	2.176	2.592
74	1.824	1.936	2.256	2.592
	$\bar{q}$ =1.607 SD = 0.078 CV = 0.0489	$\bar{q}$ =1.829 SD = 0.0585 CV = 0.0319	$\bar{q}$ =2.0017 SD = 0.0708 CV = 0.03538	$\bar{q}$ =2.4608 SD = 0.0684 CV = 0.0278
Average CV=0.036				

Table 2.5: Discharges, CV and SD of the drippers under different pressure head at 3 l/h nominal discharge

Sl. No.	Discharges at 2m pressure head, l/h	Discharges at 2.5m pressure head, l/h	Discharges at 3.5m pressure head, l/h	Discharges at 7.5m pressure head, l/h
1	1.92	2.488	2.856	3.504
2	2.224	2.584	2.8848	3.504
3	2.304	2.584	2.8848	3.552
4	2.304	2.584	2.904	3.6
5	2.352	2.584	2.904	3.6
6	2.352	2.584	2.904	3.6
7	2.352	2.632	2.904	3.6
8	2.352	2.632	2.904	3.6
9	2.352	2.632	2.904	3.6
10	2.4	2.632	2.904	3.6
11	2.4	2.632	2.904	3.6288
12	2.4	2.632	2.904	3.6288
13	2.4	2.632	2.904	3.6288
14	2.4	2.632	2.904	3.6288
15	2.4	2.6704	2.9232	3.6288
16	2.4	2.6704	2.9232	3.648
17	2.4	2.68	2.9232	3.648
18	2.4	2.68	2.9232	3.648

19	2.4	2.68	2.9232	3.648
20	2.4	2.68	2.9328	3.648
21	2.4	2.68	2.9424	3.648
22	2.4	2.68	2.952	3.648
23	2.4192	2.68	2.952	3.648
24	2.4192	2.68	2.952	3.648
25	2.4192	2.68	2.952	3.648
26	2.448	2.68	2.952	3.648
27	2.448	2.68	2.952	3.648
28	2.448	2.68	2.952	3.648
29	2.448	2.68	2.952	3.648
30	2.448	2.68	2.952	3.648
31	2.448	2.6992	2.952	3.648
32	2.448	2.6992	2.952	3.648
33	2.448	2.6992	2.952	3.648
34	2.448	2.6992	2.952	3.648
35	2.448	2.6992	2.952	3.648
36	2.448	2.7184	2.952	3.648
37	2.448	2.728	2.952	3.648
38	2.496	2.728	2.952	3.648
39	2.496	2.728	2.952	3.6672
40	2.496	2.728	2.952	3.6672
41	2.496	2.728	2.952	3.6672
42	2.496	2.728	2.952	3.696
43	2.496	2.728	2.952	3.696
44	2.496	2.728	2.952	3.696
45	2.496	2.728	2.952	3.696
46	2.496	2.728	2.952	3.696
47	2.496	2.7568	2.952	3.696
48	2.496	2.776	2.952	3.696
49	2.544	2.776	2.952	3.696
50	2.544	2.776	2.9616	3.696
51	2.544	2.776	2.9616	3.696
52	2.544	2.776	2.9712	3.696
53	2.544	2.776	2.9808	3.696
54	2.544	2.776	2.9808	3.7248
55	2.544	2.776	3	3.744
56	2.544	2.776	3	3.744
57	2.544	2.776	3	3.744

58	2.592	2.7952	3	3.744
59	2.592	2.7952	3	3.792
60	2.592	2.824	3	3.792
61	2.592	2.824	3.048	3.792
62	2.64	2.824	3.048	3.792
63	2.64	2.824	3.096	3.8304
64	2.64	2.872	3.096	3.84
	$\bar{q}$ =2.4529 SD = 0.109 CV = 0.0446	$\bar{q}$ =2.7046 SD = 0.0728 CV = 0.0269	$\bar{q}$ =2.95 SD = 0.0444 CV = 0.015	$\bar{q}$ =3.667 SD = 0.0654 CV = 0.0178
Average CV = 0.026				

Table 2.6: Discharges, CV and SD of the drippers under different pressure head at 5 l/h nominal discharge

Sl. No.	Discharges at 2m pressure head, l/h	Discharges at 2.5m pressure head, l/h	Discharges at 3.5m pressure head, l/h	Discharges at 7.5m pressure head, l/h
1	3.888	4.16	4.7904	5.52
2	3.984	4.32	4.7904	5.5552
3	3.984	4.3584	4.8	6.24
4	3.984	4.3584	4.8192	6.24
5	3.984	4.3584	4.896	6.24
6	3.984	4.368	4.896	6.288
7	3.984	4.368	4.896	6.2976
8	3.984	4.368	4.896	6.3168
9	3.984	4.368	4.896	6.336
10	3.984	4.368	4.896	6.336
11	3.984	4.368	4.896	6.336
12	3.984	4.368	4.896	6.3552
13	4.032	4.3776	4.896	6.384
14	4.032	4.3968	4.896	6.384
15	4.032	4.3968	4.896	6.384
16	4.032	4.3968	4.9152	6.384
17	4.032	4.3968	4.9152	6.384
18	4.032	4.3968	4.9152	6.384
19	4.032	4.416	4.9152	6.384
20	4.08	4.416	4.9152	6.384
21	4.08	4.416	4.9152	6.4128
22	4.08	4.416	4.944	6.432
23	4.08	4.416	4.944	6.432

24	4.08	4.416	4.944	6.432
25	4.08	4.416	4.944	6.432
26	4.08	4.416	4.944	6.432
27	4.08	4.416	4.944	6.432
28	4.08	4.416	4.944	6.432
29	4.08	4.416	4.944	6.432
30	4.08	4.416	4.944	6.432
31	4.08	4.4352	4.944	6.432
32	4.08	4.4352	4.944	6.432
33	4.08	4.4352	4.944	6.4512
34	4.08	4.464	4.944	6.4512
35	4.08	4.464	4.944	6.48
36	4.08	4.464	4.944	6.48
37	4.08	4.464	4.944	6.48
38	4.08	4.464	4.944	6.48
39	4.08	4.464	4.944	6.48
40	4.1088	4.464	4.944	6.48
41	4.128	4.464	4.944	6.48
42	4.128	4.464	4.944	6.48
43	4.128	4.464	4.944	6.48
44	4.128	4.464	4.944	6.48
45	4.128	4.464	4.944	6.48
46	4.128	4.464	4.944	6.48
47	4.128	4.464	4.9632	6.528
48	4.128	4.4736	4.9728	6.528
49	4.128	4.4928	4.9728	6.528
50	4.128	4.4928	4.9728	6.528
51	4.128	4.4928	4.9728	6.528
52	4.128	4.4928	4.992	6.528
53	4.128	4.4928	4.992	6.576
54	4.132	4.4928	4.992	6.576
55	4.176	4.4928	4.992	6.576
56	4.176	4.4928	4.992	6.576
57	4.176	4.5024	4.992	6.576
58	4.176	4.512	4.992	6.576
59	4.176	4.512	4.992	6.576
60	4.176	4.512	4.992	6.6048
61	4.176	4.512	4.992	6.6048

62	4.176	4.512	5.0112	6.6048
63	4.176	4.512	5.0112	6.624
64	4.224	4.512	5.04	6.624
65	4.224	4.512	5.04	6.624
66	4.224	4.512	5.04	6.624
67	4.224	4.512	5.04	6.624
68	4.224	4.512	5.04	6.624
69	4.224	4.5312	5.04	6.624
70	4.224	4.5312	5.04	6.624
71	4.224	4.5312	5.04	6.672
72	4.272	4.5312	5.04	6.672
73	4.272	4.56	5.04	6.672
74	4.32	4.56	5.088	6.672
75	4.32	4.56	5.088	6.672
76	4.32	4.56	5.088	6.7008
77	4.32	4.56	5.088	6.7008
78	4.32	4.56	5.136	6.72
79	4.32	4.56	5.136	6.72
80	4.32	4.608	5.184	6.7392
81	4.32	4.704	5.184	6.768
82	4.8	4.8	5.28	6.768
	$\bar{q} = 4.13$ SD = 0.1256 CV = 0.03	$\bar{q} = 4.463$ SD = 0.0845 CV = 0.0189	$\bar{q} = 4.97$ SD = 0.08367 CV = 0.0168	$\bar{q} = 6.48$ SD = 0.197 CV = 0.03
Average CV = 0.062				

Irrigation Uniformity and Efficiency

The discharges from the distributors depend on:

i. Designed distributors characteristics
ii. Standard of manufacture
iii. Friction losses in the pipe network
iv. Elevation fluctuation of the field
v. Number of clogging or partially clogged distributor within the system
vi. Variation in the water temperature within the system.

So far, there is no analytical means to deal with all the above items in drip system design. However, in a good irrigation system it is desired to ensure sufficient water to least-watered plant. Therefore, the relationship between the minimum discharges of the distributors to the average discharge is an important factor. The degree of emitter flow variation in a lateral can be expressed in different ways, such as:

i. $Eu = 1 - \frac{\overline{\nabla q}}{\overline{q}}$ (2.35)

ii. $E_u = 1 - \frac{q_{\underline{min}}}{\overline{q}}$, Keller and Bliesner (1990) (2.36)

iii. $q_{var} = \frac{q_{max} - q_{min}}{q_{max}}$, Wu and Gitlin (1974) (2.37)

Where, E_u = uniformity coefficient of the emitter flows

$\overline{\nabla q}$ = mean of absolute deviation from the mean emitter flow

$\overline{q}$ = mean of emitter flow

$q_{\underline{min}}$ = average discharge of the lowest one fourth of the emitters

q_{var} = emitter flow variation

q_{min} = minimum emitter flow

q_{max} = maximum emitter flow

The quantitative expression of Eq.2.37 is very simple because it requires only the maximum and minimum flow. Since the pressure flow variation vis-à-vis the emitter flow variation are smooth curves (Fig. 2.2-2.4), the relationship between the emitter flow variation, q_{var}, and the uniformity coefficient (Eq.2.36) can be obtained as shown in Fig. 2.5. A uniformity coefficient about 98% equals an emitter flow variation of 10% and a uniformity variation coefficient of 95% equals an emitter flow variation of 20%.

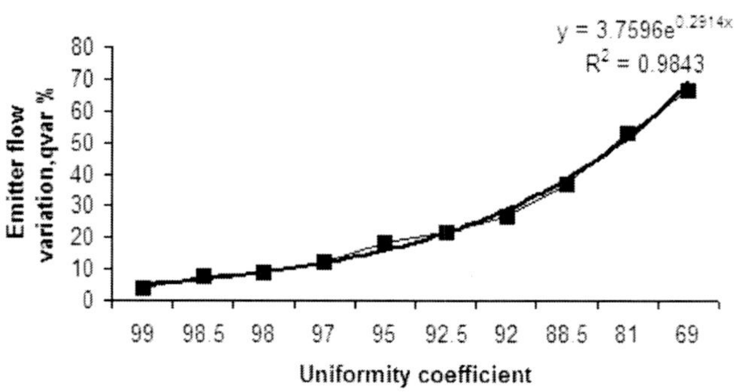

Figure 2.5: Relationship between emitter flow variation and uniformity coefficient

The pressure and emitter flow variation are related by the x-value ($q = K_d H^x$) as shown in Eq.5.21.

$$q_{var} = 1 - (1 - H_{var})^x \tag{2.38}$$

$$\text{and, } H_{var} = \frac{H_{max} - H_{min}}{H_{max}} \tag{2.39}$$

Keller and Karmeli (1971) defined the designed E_u (emission uniformity) as "the manufacturer's discharge ratio, adjusted for the number of distributors per plant and expressed as a percentage, multiplied by the ratio of the absolute minimum, determined from the nominal rate versus head curve to the average distributor discharge rate and expressed as the following equation:

$$Eu = 100\frac{q_{min}}{\overline{q}}M_r f(e) \text{ with } M_r f(e) = 1 - \frac{1.27CV}{\sqrt{e}} \tag{2.40}$$

Where, q_{min}=the minimum discharge rate of the distributors determined with the minimum pressure, within the applicable range, causing the nominal relationship q & h.

$\overline{q}$ = the average discharge rate of all the pressure

M_r= the manufacturer's discharge ratio

f(e) = the adjustment factor for number of distributors per plant.

The manufacturer's discharge ratio is the average of the low ¼ to the average discharge rate of a test sample of distributors operated at a reference pressure head and estimated from the CV. The average of the low ¼ is taken as the practical minimum.

The Eq.2.40 may be rewritten as

$$E_u = 100\left(1 - \frac{1.27CV}{\sqrt{e}}\right)\frac{q_{min}}{\overline{q}} \tag{2.41}$$

Example 2.6 The standard deviation of discharges and average discharge rate are 5% and 5.05 l/h respectively of a set of distributors. What is the maximum and minimum flow through the distributors? What is the emission uniformity if 6-oulet distributors are used?

Solution: Manufacturer's coefficient of variation, $CV = \frac{\sigma}{\overline{q}} = \frac{0.05}{5.05} = 9.9\text{x}10^{-3}$

$$q_{max\,or\,min} = \left\{1 \pm \frac{3\sigma}{\overline{q}}\right\}\overline{q} = \{1 \pm 3CV\}\overline{q}$$

$$q_{max} = \left\{1 + \frac{3\sigma}{q}\right\} = \{1 + 9.9\text{x}10^{-3\text{x}3}\}5.05 = \{1 + 0.00297\}5.05 = 5.201/h$$

$$q_{min} = \{1 - 3CV\}\overline{q} = \{1 - 0.0297\}5.05 = 4.901/h$$

$$\therefore E_u = 100\frac{q_{min}}{\bar{q}}\left(1-\frac{1.27CV}{\sqrt{e}}\right)$$

$$= 100\frac{4.90}{5.05}\left(1-\frac{1.27x9.9x10^{-3}}{\sqrt{6}}\right)$$

$$= 97.03(1-5.13x10^{-3})$$

$$= 96.53\%$$

Questions & Problems

2.1. Prove that $h_l = \dfrac{128nlq}{\pi d^4 g}$ in drip laminar flow.

2.2. How the roughness of pipes influence on friction factor, f?

2.3. What is a distributor? What are the characteristics the distributors should have?

2.4. What is manufacturers' coefficient of variation? How is it important to uniformity of discharge?

2.5. How do you calculate the CV? What is the important physical significance of CV?

2.6. What are the factors on which the discharge of distributors depends? State the formulae, which are commonly used to express the uniformity coefficient of the distributors.

2.7. How Keller and Karmeli defined the emission uniformity?

2.8. Determine the characteristics flow, which takes place in a lateral pipe at a rate of 500lit/hr. The diameter of lateral and the kinematics viscosity of water are 12mm and $5.5x10^{-7}m^2/sec$ respectively. What is the value of friction coefficient, f, if the lateral is a relatively rough pipe?

2.9. Determine the energy loss in a drip partially laminar flow with Reynolds number 3500. The length of the pipe 25m, diameter of pipe 15mm and velocity of water 50cm/s.

2.10. Develop the q (discharge)-H (head) relationship form the following data:

q(l/h)	1.12	1.52	1.85	2.15	2.40	2.6	2.83	3.10	3.22
H (m)	2	4	6	8	10	12	14	16	18

2.11. The following discharges (l/h) were recorded in the distributors at the nominal head of 10m:

4.01, 4. 03, 4.19, 4.15, 4.05, 3.97, 3.92, 3.96, 4.08, 3.90, 4.14, 4.10, 3.92, 3.94, 4.07, 4.20,4.13, 4.09, 3.91,3.99, 4.07, 3.89, 4.17, 4.11, 4.13, 3.93, 4.17 and 3.95.

Determine the coefficient of manufacturing variation for this set of emitters. Find the $q_{\max}$, $q_{\min}$ and E_u.

2.12. What is the maximum and minimum flow through distributors when standard deviation of discharge and average discharge are 5% and 4.05l/h respectively. What is the coefficient of variation (CV) of discharges through the distributors?.

Ans: q_{min} = 3.85l/h, q_{max} = 2l/h, CV= 0.0123

2.13. The discharge of distributors are characterized by $q = 0.65H^{0.8}$. The minimum and average pressure head in a lateral are 9 and 10m respectively. Determine the CV.

Ans: CV = 0.027

2.14. The standard deviation of discharge and average discharge rate of a set of distributors are 4.5% and 6.5l/h respectively. What is the minimum uniformity of 4-outlet distributors are used?

Ans: 97.57%

2.15. Write True or False of the following:

1. In laminar flow friction coefficient is independent to relative roughness of pipe.
2. Reynolds number is more than 5000 in fully turbulent flow
3. Reynolds number is unimportant when turbulent flow occurs on rough surface.
4. Drip system is standardized at water temperature of 25^{0}C.
5. Discharge coefficient C_d depends on the characteristics of orifice or nozzle.
6. The variation in distributor diameter causes more variation to discharge in laminar flow than turbulent flow.
7. Manufacturer's coefficient of variation of distributor discharge should be within 0.1 for good distributor.
8. More the number of distributors in a plant more the manufacturer's coefficient of variation of distributor discharges.

9. Approximately 68% of the discharge rates of the distributors fall within $(1 \pm CV)\bar{q}$.
10. In a fully compensating distributor the discharge to 0.5 power of pressure head.

Ans. 1.True 2. False 3. True 4. False 5. True 6. False 7. False 8. False 9. True 10. False.

2.16. Select the appropriate answer from the following.

1. Reynolds number in circular pipe may be expressed as:

a) $R_e = \frac{vd}{\nu}$ b) $R_e = \frac{vd^2}{\nu}$

c) $R_e = \frac{v^2 d}{\nu}$ d) $R_e = \frac{vd}{g\nu}$

2. The loss of energy in pipe is

a) Directly proportional to diameter

b) Directly proportional to square of diameter

c) Inversely proportional to the diameter

d) Inversely proportional to the square of diameter

3. Unstable flow starts when Reynolds number exceeds

a) 1000 b) 1500

c) 2000 d) 4000

4. Normal temperature of water in drip is assumed

a) 5^0C b) 10^0C

c) 15^0C d) 20^0C

5. The discharge-head relationship of a distributor is $q = K_d H^x$. In fully turbulent flow the value of exponent x is

a) 0.5 b) 0.75

c) 0.85 d) 1.0

6. Water flow at a rate of 0.15l/s through a lateral pipe of 10 mm diameter. If kinematics viscosity of water is 5.6x2$10^{-7}$$m^2$/s, the Reynolds number is

a) 33928 b) 16964

c) 25446 d) 67856

7. The dimension of kinematic viscosity is

a) L^2T^{-11} b) LT^{-1}

c) LT d) L^2T

8. In a flow regime the Renolds number is 1750. Flow regime is said to be

   a) Laminar b) Unstable

   c) Partially turbulent d) Turbulent

9. In a laminar flow the Reynolds number (R_e) is 1750. The coefficient of friction (f) is about

   a) 0.03 b) 0.04

   c) 0.05 d) 0.06

10. The coefficient of friction (f) in a turbulent flow is 0.02. The Reynolds number is about

    a) 18791 b) 37581

    c) 46977 d) 62636

11. At fully turbulent flow a distributor discharges 4l/h at 10m pressure. The discharge coefficient (K_d) of the distributor is about

    a) 0.86 b) 1.16

    c) 1.26 d) 1.46

12. A set of distributors of average discharges 4l/h and manufacturer's coefficient of variation is 0.03. The average discharge of low one-fourth of the discharge rates is about

    a) 3.85l/h b) 3.95l/h

    c) 4.12l/h d) 4.36l/h

13. The manufacturer's coefficient of variation of a set of distributors is 4%; 6 numbers of which are used in a plant. The effective coefficient of variation of the distributors is about

    a) 0.04 b) 0.03

    c) 0.02 d) 0.01

14. Flow takes place through a 15mm relatively rough pipe of roughness dimension 0.5mm. The friction coefficient is

    a) 0.05 b) 0.06

    c) 0.07 d) 0.08

15. In an unstable flow the Reynolds number is 3000. The friction coefficient is more close to:

    a) 0.025 b) 0.03

    c) 0.035 c) 0.04

Ans. 1. (a) 2. (c) 3. (c), 4. (d), 5. (a) 6. (a) 7. (a) 8. (a) 9. (c) 10. (d) 11. (c) 12. (a) 13. (d) 14. (b) 15. (b)

References

FAO(1980).Vermeiren, I and Jobling, G.A. Irrigation and Drainage Paper 36, FAO.Rome

Karmeli, D. and Keller, J. (1975). Trickle Irrigation Design. Rain Bird Sprinkler Manufacturing Corp., Glendora, California. pp.133

Schwartzman, M and B. Zur (1985). Emitter spacing and geometry of wetted soil volume. J.Irri. Drainage Engr, ASCE, 112 (3):242-253.

Sivanappan, R.K. *et al.* (1987). Drip Irrigation. National Committee for Use of Plastic in Agriculture. Keerthi Publishing House.

3

Drip Design Procedure

Crop Water Requirements

In designing an irrigation system, the primary objective is to know the water requirement of crops. Water requirement of crops are determined either by field cultivation or by estimation through commonly used empirical formulae such as Blaney Criddle, radiation, Penman or pan evaporation method.

Definitions

Crop water requirement or evapotranspiration (ET_{crop}): It may be defined as the rate of evapotranspiration of a disease free crop growing in a field of not less than one hectare under adequate fertility and water supply so that full productive potential can be achieved in prevailing environment. Evapotranspiration requirement of a crop refers the ET_{crop} and expressed in mm/day.

Reference crop evapotranspiration (ET_0): It is the rate of evapotranspiration from an extended surface of 8 to 15cm tall green grass of uniform height, actively growing, completely shading the ground and not short of water. ET_0 may be computed by using the empirical formulae and meteorological data for the specific period.

Crop coefficient (K_c): It is the ratio of crop evapotranspiration, ET_{crop}, and the reference crop evapotranspiration, ET_0, when both apply to large field under optimum growth condition.

Ground cover: Drip irrigation system is mainly used to irrigate orchard or row crops. At the young stage of crops the most of fields remain unshaded. It is more usual to wide spaced crops. During this period there is little loss of water in the form of evaporation and transpiration from the unshaded portion whereas it occur considerably in surface or sprinkler irrigation. Thus, water requirement in crops in conventional method includes the non-beneficial evaporation or transpiration from the unshaded or uncovered areas in the field. This is taken into account in calculating the water requirement of crop in drip irrigation and is modified by introducing the reduction factor called ground coverage factor (k_r). The method of accurate estimation of k_r is still to be developed. However, the following relationships may be used for approximate estimation in which the ground cover, GC, is the fraction of the total surface area actually covered by the foliage of the plants when viewed from directly above (FAO, 1980).

i. Keller and Karmeli (1974)

$$K_r = \frac{GC}{0.85}, \text{ whichever is the smallest} \qquad (3.1)$$

ii. Freeman and Gazzoli

$$K_r = GC + \frac{1}{2}(1 - GC) \qquad (3.2)$$

iii. Decroix

$$K_r = (0.10 + GC) \text{ or } 1, \text{ whichever is the smallest} \qquad (3.3)$$

The value of 0.10 includes the oasis effect, which is very important when the coverage is small.

With the consideration of above terms and relationships, the water requirement of crop in drip irrigation is

$$ET_{crop} = K_c \times K_r \times ET_0 \qquad (3.4)$$

We know the pan evaporation (E_{pan}) is very close to ET_0. Therefore, ET_0 often estimated by a multiplication factor called pan factor (K_p) to E_{pan}. Therefore, the Eq.3.4 becomes,

$$ET_{crop} = K_p \times K_c \times K_r \times E_{pan} \qquad (3.5)$$

Example 3.1 The following are the pan evaporations (E_{pan}), reference crop evapotranspirations (ET_0) and crop coefficients (k_c) of an arbitrary location and crop.

	Jan	Feb	Mar	April	May	June	July	Aug	Sep	Oct	Nov	Dec
E_{pan}, mm/day	2.5	2.6	4.1	4.3	5.2	5.2	5.1	4.9	4.8	3.5	2.8	2.7
ET_0, mm/day	1.75	1.90	3.0	3.25	4.1	4.2	3.9	3.7	3.3	2.8	1.9	1.8
k_c	0.6	0.6	0.7	0.75	0.8	0.8	0.9	0.9	0.8	0.7	0.6	0.6

Determine the K_p values and water requirement of crop (ET_{crop}) at different months. Assume 45% ground coverage from January to March and 60% for the remaining months.

Solution: Using Decroix assumption,

$$K_r = 0.1 + GC$$

$= 0.1 + 0.45$

$= 0.55$, when GC is 45%

& $k_r = 0.1 + 0.6 = 0.7$, when GC is 60%.

ET_{crop} in any month $= K_p \times K_c \times K_r \times ET_{pan}$

For the month of January, $K_p = \frac{ET_0}{E_{pan}} = \frac{1.75}{2.5} = 0.7$

$ET_{crop} = 0.7 \times 0.6 \times 0.55 \times 2.5$

$= 0.58$ mm/day

Similarly the ET_{crop} of other months are calculated and tabulated as below.

	Jan	Feb	Mar	April	May	June	July	Aug	Sep	Oct	Nov	Dec
E_{pan}, mm/day	2.5	2.6	4.1	4.3	5.2	5.2	5.1	4.9	4.8	3.5	2.8	2.7
ET_0, mm/day	1.75	1.90	3.0	3.25	4.1	4.2	3.9	3.7	3.3	2.8	1.9	1.8
K_p	0.7	0.73	0.73	0.76	0.79	0.81	0.76	0.76	0.69	0.8	0.68	0.67
K_c	0.6	0.6	0.7	0.75	0.8	0.8	0.9	0.9	0.8	0.7	0.6	0.6
K_r	0.55	0.55	0.55	0.7	0.7	0.7	0.7	0.7	0.7	0.7	0.7	0.7
ET_{crop}, mm/day	0.58	0.63	1.15	1.72	2.3	2.36	2.44	2.35	1.85	1.37	0.8	0.76

Example 3.2 The following information were obtained in a research of papaya cultivation under drip irrigation at Gayeshpur, West Bengal (Biswas *et al.*, 1999):

The daily requirement of a plant was determined by, $V = E_{pan} \times K_p \times K_c \times A$

Where, V = volume of water applied to each plant, l/day

E_{pan} = pan evaporation x 1.0, 0.8 and 0.6 at the irrigation level of I_1, I_2 and I_3 respectively, mm/day

K_c = crop factor (assumed 0.8, 1.0 and 1.2 at 4-6, 7-9 and 10-12 months respectively of the crop age)

K_p = pan factor = 0.8

A = area under a plant = 2m × 2m

Yield = 36225, 40200, 38550 and 31200 kg/ha for I_1, I_2, I_3 and conventional method of irrigation respectively

a. Determine the volume of water applied at each plant at different irrigation levels for a pan evaporation of 5mm in a day at 5th month of crop age.

b. If the cumulative pan evaporation (CPE) during the period except in the rainy days for which irrigation was required was 240mm, find the total depth of water applied in each irrigation level.

c. Determine the irrigation water use efficiency in different irrigation treatment and percent increase in yield over conventional method.

Solution

a. $V = E_{pan} \text{ x } K_p \text{ x } K_c \text{ x } A$

$$\therefore I_1 = 5\text{mm x } 1.0 \text{ x } 0.8 \text{x } 0.8 \text{x } 4\text{m}^2$$

$$= \frac{5}{1000} \text{x } 4 = 0.02\text{m}^3 = 12.81/\text{day}$$

$$I_1 = 5\text{mm x } 0.8\text{x } 0.8\text{x } 0.8\text{x } 4\text{m}^2$$

$$= 10.241/\text{day}$$

$I_3 = 5\text{mm x } 0.6 \text{ x } 0.8 \text{ x } 0.8 \text{ x } 4\text{m}^2$

= 7.68 l/day

b. Depth of water applied at any irrigation level,

$I = E_{pan} \text{ x } K_p \text{ x } K_c$

The crop coefficient is taken 1.0 as the average of the crop coefficients of 0.8, 1.0 and 1.2 during the period.

Therefore, I_1 = 240 mm x 1.0 x 0.8 x 1.0

= 19.2 cm

I_2 = 240 mm x 0.8 x 0.8 x 1.0

= 15.36 cm

I_3 = 240 mm x 0.6 x 0.8 x 1.0

= 11.52 cm

c. Irrigation water use efficiency (WUE) at irrigation level,

$$I_1 \frac{\text{Yield / ha}}{\text{depth of water applied}}$$

$$= \frac{36,225\text{kg / ha}}{19.2\text{cm}} = 1886.72\text{kg / ha} - \text{cm}$$

$$I_2 = \frac{40,200}{15.36} = 2617.19\text{kg / ha} - \text{cm}$$

$$I_2 = \frac{38,550}{11.52} = 3346.35\text{kg / ha} - \text{cm}$$

WUE in conventional method $= \frac{31,200}{24} = 1300 kg\ /\ ha\ \text{-}\ cm$

Percent increase in yield over conventional methods at irrigation levels

$$I_1 = \frac{36225 - 31200}{31200} \text{x}100 = 16.11$$

$$I_2 = \frac{40200 - 31200}{31200} \text{x}100 = 28.85$$

$$I_3 = \frac{38550 - 31200}{31200} \text{x}100 = 23.56$$

Irrigation water requirement (IR): It is the portion of water requirement of a crop or a predetermined portion of it, which is supplied through irrigation. If the water requirement of a crop is solely satisfied through irrigation then water requirement of crop and irrigation water requirement is same. In drip irrigation, water requirement of a crop is usually a predetermined portion to conventional method.

Net irrigation requirement (IR_n): Rainfall, water in soil profile and underground seepage may contribute to water requirements of plants. Subtracting these contributions to water requirement of crop is called the net irrigation water requirement.

Gross irrigation water requirement (IR_g): In confirming the net irrigation requirement to crop some additional amount of water usually requires to minimize the unavoidable system losses. Adding this additional amount to net irrigation requirement is the gross irrigation requirement. When there is some extra amount of water is required for leaching that too is included in gross irrigation requirement.

$IR_g = IR_n/E_i + L_r$

Where, E_i = irrigation efficiency and L_r = leaching requirement.

The unavoidable system losses as stated above may be account to irrigation efficiency or overall irrigation efficiency in drip irrigation. Therefore,

$E_i = E_a.E_u$

Where, E_a is the application efficiency, which may be defined as the ratio of water stored in the plant root zone to water delivered to the field. Thus, E_a takes into consideration the deep percolation or other losses of water. E_u is the uniformity of application of water through different distributors. The value of both E_a and E_u are less than 1.

Drip irrigation system does not permit the deep percolation. It is meant for only to supply the calculated volume to satisfy the crop requirement. However, the porous or light soils, which have poor water holding capacity and high infiltration rate should be provided with some additional water to meet up the unavoidable percolation losses. Vermeiren & Jobling (1980) suggested the values of E_a for different soils are listed in Table 3.1.

Table 3.1: Values of E_a for various soils

Soil type	E_a
Coarse sand or light top soil with gravel subsoil	0.87
Sands	0.91
Silts	0.95
Loam & clays	1.00

The value of E_u entirely depends on the discharge rates of different distributors in the system. Discharge from a distributor is the function of pressure variation and discharge characteristics of the distributors. Well-designed distributor network and high perfection in manufacturing the distributors gives high value of E_u.

Some researchers suggested allowing 10 percent additional water to irrigation requirement as first approximation to minimize the requirement of leaching and unavoidable percolation. Therefore, gross irrigation requirement may be written as

$$IR_g = \frac{IR_n}{E_i} x1.10$$

Peak irrigation requirement: Peak irrigation requirement is used to calculate the pipe sizes, pump capacity, etc. Once the system is designed for certain capacity it remains constant. However, the daily requirement of water may changes. This changes is met by adjusting the duration of pumping but not by increasing or decreasing the flow rate. The determination of peak irrigation requirement may arise in two cases: one when the irrigation is the sole source to meet water requirement and another when irrigation is supplemental to nature. Therefore, gross irrigation requirement may be calculated as below:

$$IR_g = E_{pan} \times K_p \times K_c \times K_r / E_i + L_r \qquad (3.6)$$

Peak pan evaporation considered is not the absolute peak pan evaporation but a compromise between the costs of supplying the absolute peak irrigation requirement to losses in production if certain reduced amount of peak demand is supplied to the plants. Instead of taking the average of pan evaporation of exceptional hot days, the peak pan evaporation is taken as average of the month of maximum evaporations. The plant factor is also very important. There are some plants whose flowering; fruit set or other water sensitive periods may coincide the extreme days of evaporation. The orchard crops with well-established and strong root system the 'average maximum value' can satisfy the water requirement of 90 to 95 percent. The plants which are more water demanding, delicate and with small root system, the peak pan evaporation should be taken as the average daily evaporation for the 10 days of maximum evaporation.

The value of K_p is considered from the published data on agro-climatic zone basis. However, when there is no such data avoidable a value of 0.8 may be taken as first approximation as the value fluctuates around this. For many maturing crops with good water holding capacity the value of K_c and K_r may be taken as unity.

The peak irrigation requirement when the irrigation is supplemental to irrigation may be fixed on the basis of local research on each crop and soil type. In general, the shallow rooted crops on lighter soils the peak irrigation requirement will be more and it will be less in deep- rooted crops with good water holding capacity.

Day-to-day irrigation requirement: The amount of water to be irrigated in a day or per irrigation to be calculated in a normal operating system by the modification of the values K_c, K_r and E_{pan}. The K_c and K_r value will depend on the stage of the growth. The E_{pan} is taken as the average daily evaporation for the months.

Example 3.3 With the data in Example 3.1 and assuming the E_a, E_u and crop spacing as 0.85, 0.90 and 5m x 5m respectively, find the gross irrigation requirement in mm/day, l/day/plant and peak irrigation requirement (PIR).

Solution: Using the available data the values of ET_{crop} were determined in Example 3.1. With these values of ET_{crop} and introducing the E_a and E_u, the gross irrigation requirement in mm/day, l/day/plant and peak irrigation requirement (PIR) are determined as below.

	Jan	Feb	Mar	April	May	June	July	Aug	Sep	Oct	Nov	Dec
E_{pan}, mm/day	2.5	2.6	4.1	4.3	5.2	5.2	5.1	4.9	4.8	3.5	2.8	2.7
ET_0, mm/day	1.75	1.90	3.0	3.25	4.1	4.2	3.9	3.7	3.3	2.8	1.9	1.8
k_p	0.7	0.73	0.73	0.76	0.79	0.81	0.76	0.76	0.69	0.8	0.68	0.61
k_c	0.6	0.6	0.7	0.75	0.8	0.8	0.9	0.9	0.8	0.7	0.6	0.6
K_r	0.55	0.55	0.55	0.7	0.7	0.7	0.7	0.7	0.7	0.7	0.7	0.7
ET_{crop}, mm/day	0.58	0.63	1.15	1.72	2.3	2.36	2.44	2.35	1.85	1.37	0.8	0.69
E_a	---------------------------- 0.85 --											
E_u	---------------------------- 0.90 --											
IR_g mm/day	0.76	0.82	1.58		2.25	3.00	3.08	3.19	3.07	2.42	1.05	0.90
IR_g lit/day/plant						75	77	79.75	76.75			
	PIR = 79.75 l/day/plant											

Water Distribution in Soils and Wetting Pattern

The shape of the distribution of water when applied from a point source in soil depends mainly on soil characteristics and gravity force. The soil texture, the soil horizontal and vertical permeability, capillary suction, presence or absence of impervious layers, the volume of water applied per irrigation, the rate of application and the initial moisture content influence the wetting pattern of soil.

The fine textured soil such as clay and clay loam, the capillary forces are strong and gravity force can be considered negligible. The horizontal movement may be faster than the downward. The wetting pattern usually takes the shape of a bulb (Fig. 3.1a). In light soil the capillary forces are small and the gravity force has some influence on movement of water. The downward movement is faster than horizontal, which causes a wetting pattern of more elongation to downward (Fig. 3.1b). The soils in between the fine and light soils the influence of capillary suction and gravity are almost equal Therefore, the wetting pattern will have more or less equal horizontal and vertical elongation leads to pear shape (Fig. 3.1c). However, the soils are very complicated in nature. Soil characteristics are seldom homogeneous. Therefore, it is very difficult to predict the exact shape of the wetting pattern.

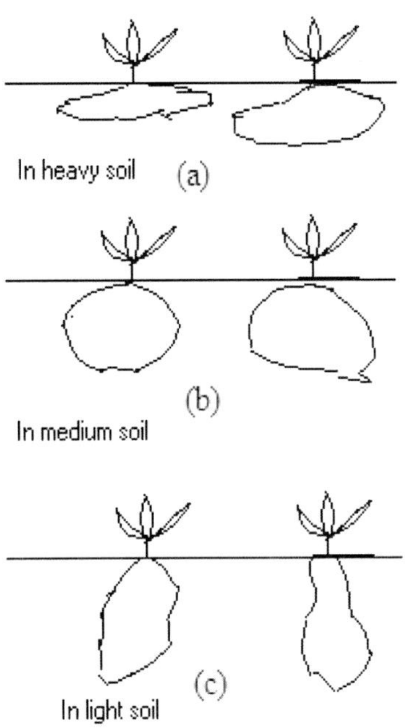

Figure 3.1: Drip water penetration pattern

Schwarzman and Zur (1985) proposed the following empirical equations correlating depth and width of the wetted soil volume to emitter discharge, saturated hydraulic conductivity of soil and volume of water in the soil volume.

$$z = K_1 (V_w)^{0.63} \left(\frac{C_s}{q}\right)^{0.45} \qquad (3.7)$$

$$w = K_2 (V_w)^{0.22} \left(\frac{C_s}{q}\right)^{-0.17} \qquad (3.8)$$

Combining Eqs.3.7 and 3.8 one can get the following relationship,

$$w = K3(z)^{0.33} (q)^{0.33} (Cs)^{-0.33} \qquad (3.9)$$

Where, z = vertical distance to wetting front, m

w = wetted width or diameter of wetting front, m

K_1 = empirical coefficient = 29.2

V_w = volume of water applied, l

C_s = saturated hydraulic conductivity of the soil, m/s

q = emitter discharge, l/h

K_2 = empirical co-efficient = 0.031

K_3 = empirical co-efficient = 0.0094

Singh *et al.* (2000) stated the investigation on the radial and vertical movement of water front in clay and clay loam soils at Rahuri, Maharashtra using different discharge rates (2, 4, 6, 8, 10 and 12 l/h) and different irrigation schedules (daily, alternate day, two and three days interval). It was observed that wetting front movement in vertical as well as radial plane was different for different emitter discharge and irrigation schedules. The average depth at which maximum radial spread occurs found to be 40 to 60cm irrespective of emitter discharge and volume of water applied. The findings of the investigation gave the following relationships for estimating the maximum lateral and vertical movement of waterfront:

For clay loam soil, $w = 0.50q^{-0.09}v^{0.24}$ (3.10)

$z = 0.48q^{0.14}v^{0.18}$ (3.11)

For clay soil, $w = 0.50q^{-0.19}v^{0.18}$ (3.12)

$z = 0.48q^{0.17}v^{0.15}$ (3.13)

where, w = maximum lateral movement, m

z = maximum vertical movement, m

q = emitter discharge, l/h

v = volume of water added, l

There are also some graphical representations and charts to estimate the extent of wetting, which was developed by limited number of field experiments (Fig. 3.2 & 3.3) (FAO, 1980). Therefore, the values are suggested to use with caution as a first approximation (FAO, 1980). Singh *et al.* (2000) conducted the experiment with three types of soil, viz. loam, sandy loam and sand with three discharges for determining the horizontal and vertical movement of water front from a point source. The relation thus developed are stated below,

$$S_y = ct^d \quad S_y = ct^d \tag{3.14}$$

$$\frac{t}{S_x} = a + bt \tag{3.15}$$

Where, S_y = vertical advance, cm

c = a constant

t = time of application, min

d = arithmetic slope of the straight line

S_x = horizontal advance of moisture front, cm

a & b = constant

Table 3.2: Vertical advances (S_y) and elapsed time (t) for different soils and flow rate

Type of soil	Discharge rate (lph)	Observed vertical distance of water (cm)	Equation	Correlation coefficient
Sandy	1.4	34	$Sy = 2.487t^{0.3646}$	0.998*
loam	1.8	40	$Sy = 2.297t^{0.4142}$	0.998*
	2.8	48.5	$Sy = 1.723t^{0.5147}$	0.997*
Loam	1.4	30	$Sy = 2.4876t^{0.3495}$	0.997*
	1.8	34	$Sy = 2.1915t^{0.3988}$	0.995*
	2.8	40	$Sy = 1.6834t^{0.49}$	0.999*
Sand	1.4	48	$Sy = 2.075t^{0.439}$	0.995*
	1.8	58	$Sy = 1.792t^{0.505}$	0.999*
	2.8	75	$Sy = 1.725t^{0.5806}$	0.995*

* Significance at 1% level

Source: Sigh *et al.* (2000)

Table 3.3: Empirical equations relating horizontal advance (S_x) and elapsed time (t) for different soils and flow rates

Type of soil	Discharge rate (Iph)	Observed horizontal distance of water (cm)	Equation	Correlation coefficient
Sandy	1.4	30.5	t/S_x =3.387 + 0.0305?	0.999*
loam	1.8	34	t/S_r =1.999 + 0.0276?	0.999*
	2.8	39	?/S_v =1.892 + 0.0227?	0.999*
Loam	1.4	24	t/S_x = 2.488 + 0.0266?	0.999*
	1.8	28	t/S_x =1.961 + 0.025?	0.998*
	2.8	29	t/S_s =1.714 + 0.0193?	
Sand	1.4	35	?/X =3.13 + 0.0395?	0.999*
	1.8	38	?/S_v = 2.895387 + 0.0355?	0.999*
	2.8	45	t/S, =1.463 + 0.0325?	0.999*

* Significance at 1% level

Source: Sigh *et al.* (2000)

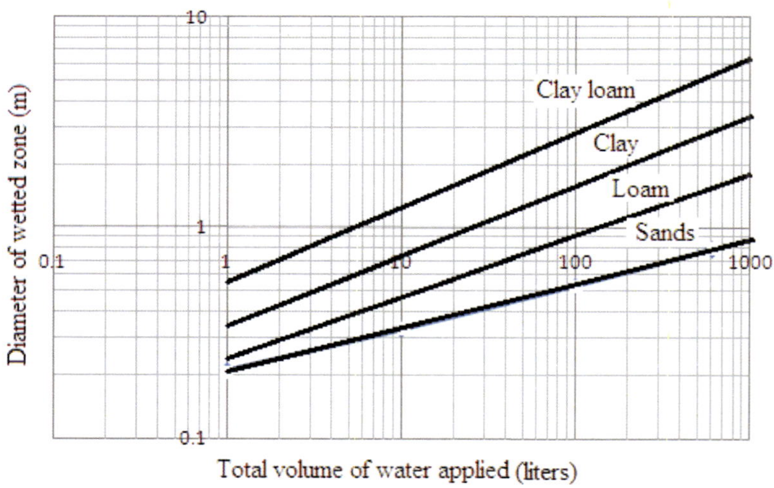

Courtesy: FAO(1980)

Figure 3.2: Approximate guide for estimating the diameter of wetting

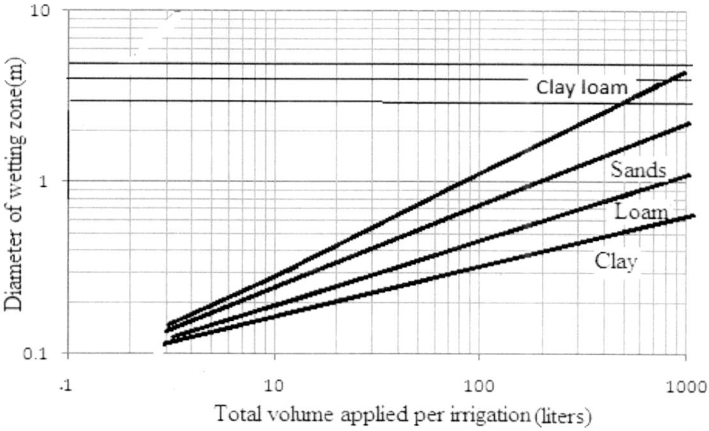

Courtesy: FAO(1980)

Figure 3.3: Approximate guide for estimating the diameter of wetting

Example 3.4 Drip irrigation is done from a point source at a rate of 2 l/h and 5 l/plant/day. If the saturated hydraulic conductivity of the soil is 1.2m/day, determine the width and depth of wetting.

Solution: Using Schwarzman & Zur (1985) [Eq.3.7 & 3.8)],

$$\text{Depth of wetting, } z = K_1 (V_w)^{0.63} \left(\frac{C_s}{q}\right)^{0.45}$$

$$= 29.2x50.63\left(\frac{1.2}{24x3600x2}\right)0.45$$

$$= -29.2x2.76x4.77x10^{-3}$$

$$= 0.38m$$

Width of wetting,

$$w = K_2 (V_w)^{0.22} \left(\frac{C_s}{q}\right)^{-0.17}$$

$$= 0.031x5^{0.22}\left(\frac{1.2}{24x3600x2}\right)^{-0.17}$$

$$= 0.031x1.42x\,7. = 53$$

$$= 0.33m$$

Example 3.5 In continuation of Example 5.4 and assuming 5, 10, 15, 20, 25 and 30 l/plant/day water application, develop the relation between water front advance and volume of water application.

Solution

Using Schwarzman & Zur(1985) equation the depths of wetting (z) and the widths of wetting (w) calculated are listed below.

Volume of water application, l/plant/day	Water front advance	
	Vertical (z), m	Horizontal (w), m
5	0.38	0.33
10	0.59	0.39
15	0.77	0.42
20	0.92	0.45
25	1.06	0.47
30	1.19	0.49

V	z
5	0.38
10	0.59
15	0.77
20	0.92
25	1.06
30	1.19

V	w
5	0.33
10	0.39
15	0.42
20	0.45
25	0.47
30	0.49

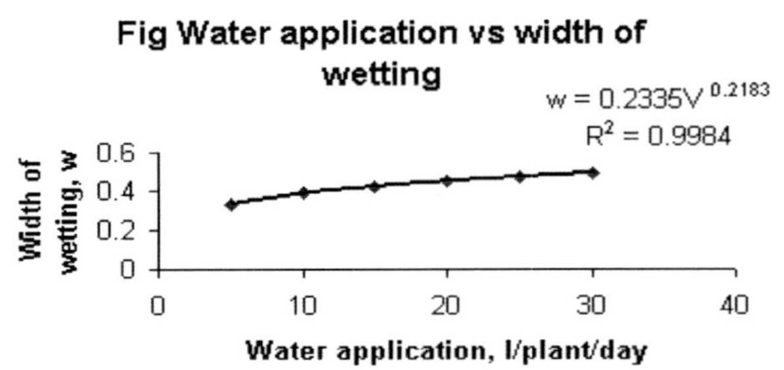

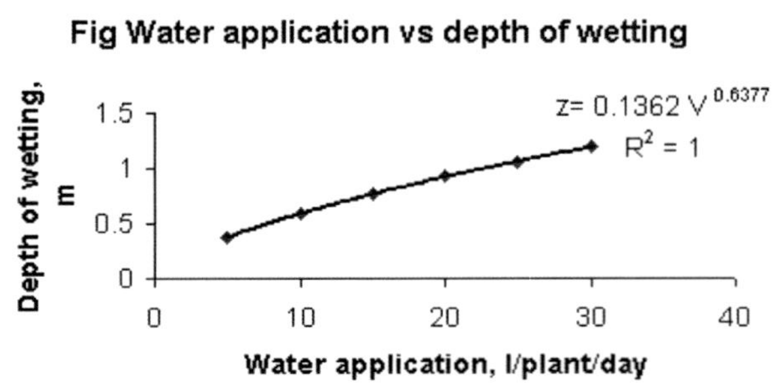

Thus, the following two equations are developed for different application volume for the application rate of 2 l/h and at saturated hydraulic conductivity of soil 1.2m/day.

Depth of wetting, $z = 0.1362V^{0.6377}$

Width of wetting, $w = 0.2335V^{0.2183}$

Example 3.6 Referring to Example 3.4, assuming point source discharge rate of 1, 2, 3, 4, 5, 7 & 10 l/h and water application of 5, 10, 15, 20 & 30l/plant/day, calculate the vertical and horizontal wetting and develop the relations between water spreading and rate of application.

Solution

Similar to Example 3.4, by using the Schwarzman & Zur(1985) equation, the depths of wetting (z) and the widths of wetting (w) calculated are listed below.

For application volume, V= 5 l/plant/day

Rate of discharge, l/h	Water front advance	
	Vertical (z), m	Horizontal (w), m
1	0.52	0.30
2	0.38	0.33
3	0.32	0.36
4	0.28	0.37
5	0.25	0.39
7	0.22	0.41
10	0.19	0.44

For V = 10 l/plant/day

Rate of discharge, l/h	Water front advance	
	Vertical (z), m	Horizontal (w), m
1	0.81	0.34
2	0.59	0.39
3	0.50	0.42
4	0.44	0.44
5	0.39	0.45
6	0.36	0.47
7	0.34	0.48
10	0.29	0.51

For 15l/plant/day

Rate of discharge, l/h	Water front advance	
	Vertical (z), m	Horizontal (w), m
1	1.05	0.38
2	0.77	0.42
3	0.64	0.45
4	0.56	0.47
5	0.51	0.50
6	0.47	0.51
7	0.44	0.52
10	0.37	0.56

For 20plant/day

Rate of discharge, l/h	Water front advance	
	Vertical (z), m	Horizontal (w), m
1	1.26	0.40
2	0.92	0.45
3	0.77	0.48
4	0.67	0.51
5	0.61	0.53
6	0.56	0.54
7	0.52	0.56
10	0.45	0.59

For 30 l/plant/day

Rate of discharge, l/h	Water front advance	
	Vertical (z), m	Horizontal (w), m
1	1.62	0.44
2	1.19	0.49
3	0.99	0.53
4	0.89	0.56
5	0.79	0.58
6	0.72	0.59
7	0.68	0.61
10	0.58	0.65

By computer analysis of the data the following charts and equations were obtained for water front advances towards vertical and horizontal at different application volume.

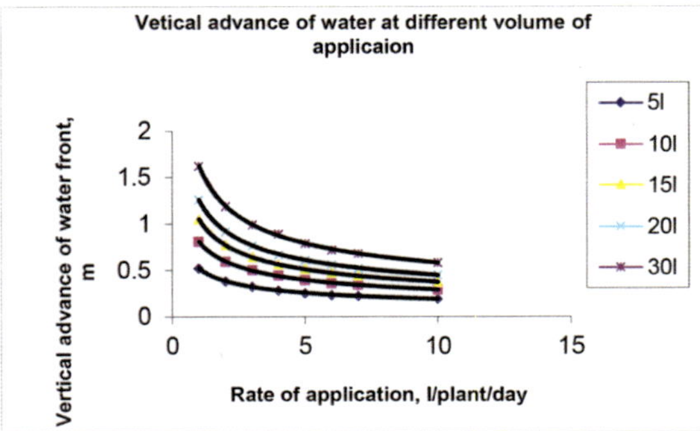

For $V = 5l,\ z = 0.5171q^{-0.4422}$

$V = 10l,\ z = 0.8098q^{-0.4474}$

$V = 15l,\ z = 1.0509q^{-0.4507}$

$V = 20l,\ z = 1.2576q^{-0.4502}$

$V = 30l,\ z = 1.6237q^{-0.4472}$

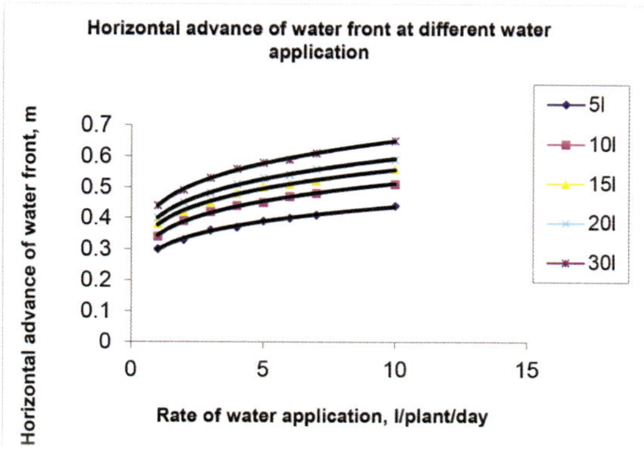

For 5l, $w = 0.2924q^{0.1663}$

10l, $w = 0.3435q^{0.1735}$

15l, $w = 0.377q^{0.1695}$

20l, $w = 0.4001q^{0.1706}$

30l, $w = 0.4391q^{0.1698}$

Example 3.7 Five liter water has been applied from point source at a rate of 1 l/h in clay and clayey loam soil. Estimate the possible width and depth of wetting of soil.

Solution: Using the empirical equation as stated by Singh *et al.* (2000)[Eq. 3.10 & 3.11],

For clay soil:

Width of wetting, $w = 0.50q^{-.19}v^{0.29}$

$= 0.50 x 1 x 1.59$

$= 0.80m$

Depth of wetting, $z = 0.48q^{0.17}v^{0.15}$

$= 0.48 x 1^{0.17} x 1.27$

$= 0.61m$

For clayey loam soil

Width of wetting, $w = 0.5 x 1^{-0.09} x 5^{0.24}$

$= 0.50 x 1^{-0.09} x 5^{0.24}$

$= 0.5 x 1 x 1.47$

$=0.74m$

Depth of wetting, z

$= 0.48q^{0.14}v^{0.18}$

$= 0.48 x 1^{0.17} x 1.39$

$= 0.64m$

Example 3.8 Referring the Example 3.7 estimate the hydraulic conductivities in clay and clayey loam soil.

Solution

The width and depth has been calculated as 0.8 and 0.61m respectively in Example 3.7.

Using Schwarzman & Zur (1985) equation:

For clay soil:

$$z = K_1 (V_w)^{0.63} \left(\frac{C_s}{q} \right)^{0.45}$$

or, $0.61 = 29.2x5^{0.63}xC_s^{-4.45} = xq^{-0.45} = 80.488C_s$

$$\text{or}, C_s^{0.45} = \frac{0,61}{80.488} = 7.57x10^{-3}$$

$C_s = 1.195x10^{-05}$m/s = 1.69m/day (vertical)

$$\text{Again}, w = K_2 (V_w)^{0.22} \left(\frac{C_s}{q} \right)^{-0.17}$$

$$0.8 = 0.031x5^{0.22} \left(\frac{C_s}{q} \right)^{-0.17}$$

$C_s^{-0.17}$(horizontal) = 18.11, $C_s = 3.99x10^{-3}$m/day (horizontal)

Selection of Number of Distributors per Plant

Proportion of area to be wetted

The percentage of area or soil volume of potential root zone which to be wetted is important in designing drip system. The percentage of wetting varies widely crop to crop. It is reported that the percentage of wetting may be as low as 25 percent (tree plants) to hundred percent to very close growing crops (vegetables). Keller and Karmeli (1974) developed a guide for estimating the wetted volume as represented in Table 3.4. This table is made with the assumption of approximately 40mm of water application per irrigation with 0.3m fairly uniform penetration of water beneath the soil, minimum percentage of wetting as 33 percent for single, straight, equally spaced lateral lines, uniformly spaced distributors for coarse, medium and light soils and for various discharges.

The Table 3.4 can be used by entering from the left column where spacing of the laterals are given. On the same line of lateral spacing the wetted percentage, P, is read in the concerned column of certain discharge and soil type. Say, the lateral spacing between the lateral is 1m. The corresponding wetted percentages are 70, 80 and 100 for medium textured soil and 1.5, 2.0 and 4.0 l/h discharge respectively.

Table 3.4: Guide for determining values of P (Percentage of soil wetted by various discharges and spacing for a single row of uniformly spaced distributors in a single line applying in a straight line applying 40mm of water per cycle over the wetted area)

Effective spacing between laterals S_l, m	**Emission point discharge**														
	Less than 1.5 lph			**2 lph**			**4 lph**			**8 lph**			**More than 12 lph**		
	Recommended spacing of emission points along the lateral for coarse, medium and fine textured soils- Se,m														
	C 0.2	M 0.5	F 0.9	C 0.3	M 0.7	F 1.0	C 0.6	M 1.0	F 1.3	C 1.0	M 1.3	F 1.7	C 1.3	M 1.6	F 2.0
	Percentage of soil wetted														
0.8	38	88	100	50	100	100	100	100	100	100	100	100	100	100	100
1.0	33	70	100	40	80	100	100	100	100	100	100	100	100	100	100
1.2	25	58	92	33	67	100	67	100	100	100	100	100	100	100	100
1.5	20	47	73	26	53	80	53	80	100	80	100	100	100	100	100
2.0	15	35	55	26	40	60	40	60	80	60	80	100	80	100	100
2.5	12	28	44	16	32	48	32	48	64	48	64	80	64	80	100
3.0	10	25	37	13	26	40	26	40	53	40	53	67	53	67	80
3.5	9	20	31	11	23	34	23	34	46	34	46	57	46	57	68
4.0	8	18	28	10	20	30	20	30	40	30	40	50	40	50	60
4.5	7	16	24	9	18	26	18	26	36	26	36	44	36	44	53
5.0	6	14	22	8	16	24	16	24	32	24	32	40	32	40	48
6.0	5	12	18	7	14	20	14	20	27	20	27	34	27	34	40

Source: FAO (1980)

Double lateral for each row of plants

In tree plants, sometime the laterals are required to be used in pairs taking the plant rows in between or the distributors used in cluster around the plant instead of using the equally spaced on the lateral (Fig.3.4). In such situation the value of P obtained from Table 3.4 may be adjusted using the following equation.

$$P = \frac{P_1 S_1 + P_2 S_2}{S_r} \qquad (3.16)$$

Where, S_l= the inner spacing (m) between the pairs of laterals, which should be taken from Table 3.4, corresponding to P=100%, the value for given emitter discharge rate, soil type and spacing.

P_1 = is taken from Table 3.4 for S_1

P_2 = is taken from Table 3.4 for S_2

S_r = the spacing between the rows of plants

S_1 = spacing between inner rows of laterals

S_2 = spacing between outer rows of laterals

S_e = spacing between the distributors in a lateral

S_t = spacing between the plants in a row

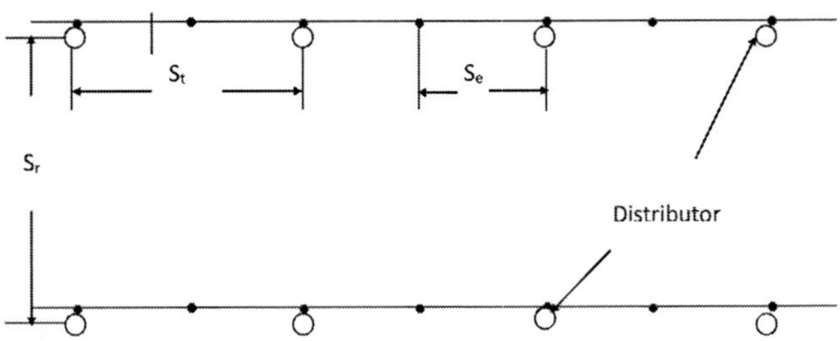

(a) Single straight lateral for each row of plants

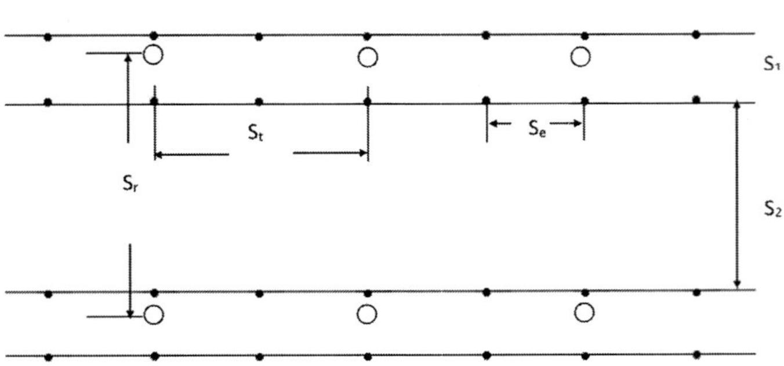

(b) Double laterals for each row of trees

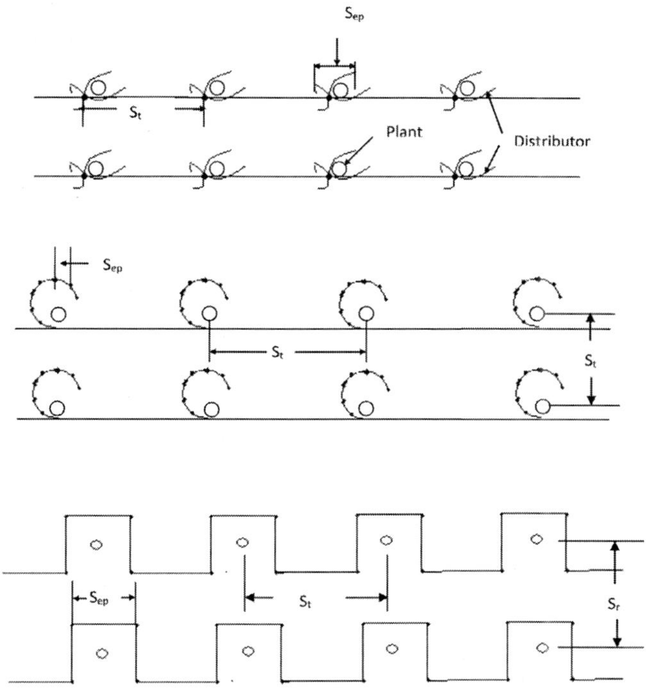

(c) Multi-exit emitters, zig-zag and pigtal layout of distributors

Figure 3.4: Distributors on laterals

Example 3.8. Design the spacing of inner rows of laterals and find the percentage of wetting of field for tree plants spaced 6mx6m in medium textured soil. The available discharge rate of distributors is 8 l/h.

Solution: Using the Table 3.4, the spacing of inner rows of laterals (S_1) is 1.5m for a value of P=100% at medium textured soil with the discharge rate of 8 l/h of the distributors. Therefore, P_1=100%.

S_2= 6m-1.5m = 4.5m

The value of P_2= 36%, from Table 3.4 for S_2=4.5m

$$P \frac{P_1S_1 + P_2S_2}{S_r}$$

$$= \frac{100x1.5 + 36x4.5}{6}$$

= 52%

Example 3.9 Tree plants are cultivated at coarse textured soil with 6.5mx6.5m spacing. The distributors fitted to the laterals pass through the plant rows and spaced 1.5m use to irrigate the plants. If the discharge rate of the distributors is 8l/h, find the percentage of area of wetting.

Solution

The inner spacing of pair of laterals, S_1=1.5m

The spacing between the rows, S_r= 6.5m

The spacing between the laterals, S_2= 6.5m-1.5m=5m

Assuming single row, equally spaced laterals with uniformly spaced distributors, P_1=80% (Using Table 3.4) for spacing (S) 1.5m, discharge rate 8l/h and coarse textured soil.

Similarly, P_2=24% for S_2=5m

The percentage of wetting,

$$P=\frac{P_1S_1+P_2S_2}{S_r}$$

$$=\frac{80x1.5+24x5}{6.5}$$

$$=\frac{240}{6.5}$$

$$=36.92\%$$

Use of multiple emissions

The shallow rooted or short duration crops suffer if water being applied from a single distributor. The anchorage is important to tree plant. Application of water to tree plant at early stage may be suitable by the single distributor. As the tree plants grow further common sense suggests that water should be applied uniformly around the plant bottom for the uniform development of root system in any direction and provides better anchorage. This may be done by increasing the number of the distributors one to two and two to few and making the distributors pointing progressively outward matching with the growth of trees from the trunk to encourage spreading of the roots. There may be different arrangement of distributors around the plant bottom. The Fig. 3.5 illustrates a few possible arrangements. The number of emission points and their spacing related by the following equation

$$\frac{P}{100}=\frac{nS_{ep}S_w}{S_tS_r} \tag{3.17}$$

Where, n= the number of emission points per tree

S_{ep}= the spacing between emission points

S_w= the width of wetted strip, which corresponds to the S_l value taken from Table 3.4, giving P=100% for the given emission discharge and soil type.

S_t= the spacing between trees in the rows

S_r= the spacing between tree rows

Example 3.10 Find out the wetted area in a tree cultivation spaced 7mx7m with 3 numbers of distributors per plant, the spacing of distributors 2m and the discharge rate 4l/h of each distributor. Assume medium textured soil.

Solution: Spacing between trees, S_t= 7m

Spacing between tree rows, S_r=7m

Number of distributors, n = 3

Spacing between emission points, S_{ep}= 2m

S_w= 1.2m from Table 3.4 for emitter discharge of 4l/h, medium textured soil and P=100%

$$\text{Now,}\frac{P}{100}=\frac{nS_{ep}S_w}{S_tS_r}$$

$$=\frac{3x2x1.2}{7x7}$$

$$+ 0.1469$$

$$\therefore P=14.69\%$$

Irrigation interval

The net irrigation water applied in irrigation can be expressed as

$$IR_n=(FC-WP)d_mZ\frac{P}{100} \tag{3.18}$$

Where, IR_n= depth of water to be applied, mm

FC= the volumetric moisture content at field capacity, mm/m

WP= the volumetric moisture content at wilting point, mm/m

d_m= the moisture depletion allowed or desired, percent

Z= root zone depth to be considered, m

P= the percent of wetted soil to total soil volume

The root zone depth varies much crop to crop. The crops also grow in wide range of soil condition. In absence of information in these regards, as the first approximation the Table 3.5 & 3.6 respectively may be used for the design purpose.

Table 3.5: Maximum/minimum values of Z for different crops

Crop	Z, m
Tomatoes	1.0-1.2
Vegetables	0.3-0.6
Citrus	1.0-1.2
Deciduous fruit	1.0-2.0
Grapes	1.0-3.0

Table 3.6: Physical properties of some soil

Soil texture	Available moisture percentage by weight			Water holding capacity by volume, mm/m
	FC	WP	Available	
Sandy	9 (6-12)	4 (2-6)	5 (4-6)	85 (70-100)
Sandy loam	14 (10-18)	6 (4-8)	8 (6-10)	120 (90-150)
Loam	22 (18-26)	10 (8-12)	12 (10-14)	170 (140-190)
Clay loam	27 (25-31)	13 (11-15)	14 (12-16)	190 (170-220)
Silty clay	31 (27-35)	15 (13-17)	16 (14-18)	210 (180-230)
Clay	35 (31-39)	17 (15-19)	18 (16-20)	250 (200-250

No design is suggested where the moisture status is maintained for maximum potential evapotranspiration. Again the crop should not go under water stress. There are some crops, which are very much sensitive to water stress. These crops are designed for the irrigation at 30% depletion of the moisture and the crops that are less sensitive to water stress are designed for 60% moisture depletion.

Evapotranspiration rate and soil moisture

Evapotranspiration occurs by overcoming the resistance to movement of water depends on the particular plant, the type of soil, soil moisture content and the evapotranspiration itself. Denmead & Shaw (1962) stated that at a lower evapotranspiration rate the movement of water through the soil is little restricted until the moisture content is very low (i.e., close to wilting point). On the other hand at high evapotranspiration rate a small reduction in soil moisture corresponding to small change in soil moisture significantly affects the transpiration rate. The inference can be made from this study that there is no field optimum irrigation interval nor any fixed suction at which to irrigate; but rather irrigation should be made at low soil moisture suction so that transpiration can occurs at the designed rate under the prevailing atmospheric condition. This means that irrigation should be frequently in hot period at low suction and long interval at considerable suction in cooler period. Demean and Shaw suggested some values of soil suction at various pan evaporation (Table 3.7).

Table 3.7: Guide to limit soil suction

Prevailing Class A pan evaporation (mm/day)	Maximum soil suction at which evapotranspiration can be maintained at 80% of maximum
2	3.5 bars
3	2.0 bars
4	1.4 bars
5	0.4 bars

Source: FAO(1980)

Irrigation at fixed deficit

The irrigation at fixed deficit determines the time of irrigation when the soil has reached to a predetermined water deficit or otherwise after the fixed amount of evaporation. That means the cumulative evaporation to be divided by PIR to find out the irrigation interval. However, pulse irrigation may be required to minimize the losses in light soil. Usually two-day consumption would be considered for designed deficit for heavy soil. This method, therefore, provides frequent irrigation during high evapotranspiration and less frequent during low evapotranspiration and using same volume of water in such irrigation to bring the soil in field condition. This situation gives advantage in automation to irrigation by connecting the pan and the sensor to the irrigation system.

Irrigation at fixed interval

In fixed irrigation interval the predetermined frequency is such that it will meet up the maximum water requirement of the plants. Of course, the irrigation should be frequent in hot atmosphere and less frequent in cool atmosphere. In light soils the frequency will be more frequent to heavy soils. In shallow rooted crops like vegetables the irrigation interval will be much closer compared to irrigation interval to deep rooted plants like trees.

The Table 3.8 suggests the irrigation interval for different crops, soils and atmospheric conditions.

Table 3.8: Suggested irrigation intervals

Climate	Soil		
	Very coarse, no water holding capacity	**Light sandy**	**Heavier loams, and clayey soils**
Hot and dry, high transpiration rate	Pulse irrigation during the day or once a day, when plants are using most water	1 day interval or 2 days when some silt or clay in soil	2 or 3 days interval in heavy soils which have poor aeration
Moderate	Pulse irrigation during the day or once a day, when plants are using most water	2 or 3 days interval	3 or 4 days interval
Cool, low transpiration rate	Pulse irrigation during the day or once a day, when plants are using most water	3 or 4 days interval (twice a week) providing there be some water holding capacity	6 to 8 days interval (once a week)

Duration of each irrigation

In general the duration of irrigation should be as long as possible, which suggests a low flow rate. In coarse soil the application rate should be kept close to consumptive rate to the purpose of avoiding application loss due to deep percolation and it should be preferably applied during the day time when consumption rate is high. The time of duration is influenced by any underestimate of crop requirements, abnormal peak requirement, breakdowns, general maintenance and slow decrease in average distributor discharge with time. For porous soils the suggested application time is 6 to 10 hours per day and 10 to 18 hours per day for soils of good water holding capacity. During the period of peak demand the duration may be increased to 20 to 22 hours per day.

Discharge per distributor or set of distributors

When the amount of water and the duration of application are selected, the discharge per distributor or group of distributors can be automatically found by the equation,

$$q_d = \frac{IR_g A}{I_d} = \frac{IR_g}{I_d} S_e S_l$$

Where, q_d = the discharge of a distributor or group of distributors

IR_g = gross irrigation requirement or the depth of water considered for the period of irrigation interval

I_d = the period of each irrigation

A= the area allocated to each plant

S_e= spacing between the distributors

S_l= spacing between the laterals.

Example 3.11 Determine the maximum interval between two consecutive drip irrigations from the following data:

PIR=6.2mm/day

FC=16%

PWP=5%

BD=1.35g/cm^3

Root zone depth=1.5m

Maximum allowable soil moisture depletion=30%

Design objective is to wet the potential root zone=40%

Solution: Volumetric moisture content of FC=16xBD=16x1.35=21.6% or 21.6 cm/m

Volumetric moisture content of FC WP=5xBD=5x1.35=6.75% or 6.75 cm/m

We know,

$$IR_n = (FC - WP)d_m Z \frac{P}{100}$$

$$= (21.6 - 6,75)\text{cm} / \text{m} \text{x} \frac{30}{100} \text{x} 1.5 \,\text{m} \text{ax} \frac{40}{100}$$

$$= \frac{14.85\text{x}1.5\text{x}1200}{100\text{x}100}$$

$$= 2.67 \text{ cm}$$

$$\text{Maximum interval} = \frac{IR_n}{PIR}$$

$$= \frac{2.67\text{cm}}{6.2\text{mm} / \text{day}}$$

$$= 4.3 \text{ days}$$

Example 3.12 Determine the type of distributors to be used, the frequency of irrigation, the spacing of distributors on the laterals, the spacing between the laterals, the number of distributors per tree and the duration of each irrigation from the following data:

Spacing of crop=5mx5m

Plant coverage=75%

Sandy soil, $E_a = 90\%$ and $E_u = 90\%$

FC = 10% by weight basis

WP = 4% by weight basis

BD = 1.5 g/cm^3

Maximum allowable soil moisture depletion = 33%

Effective root zone depth = 1.5m

ET_{crop} = 5.8 mm/day

Design objective of wetting=35 to 40% of potential root zone

Nominal discharges of the distributors available=2, 4 and 6l/h

Solution

i. Net water requirement of crop (ET_{crop}) is 5.8mm/day or 5.8mm/day x 5m x 5m=145l/day/plant

ii. Maximum net depth of water to be applied in each irrigation

$$IR_n = (FC - WP) d_m Z \frac{P}{100}$$

$$= (10x1.5 - 4x1.5) cm/m x \frac{33}{100} x1.5m x \frac{40}{100}$$

$$= \frac{9x1980}{10000} = 1.782cm$$

$$= 17.82$$

iii. The maximum interval between two consecutive irrigation

$$I_i = \frac{IR_n}{ET_{crop}} = \frac{17.82mm/irrigation}{5.8mm/day}$$

$$= 3.07 \text{ days, say } 3 \text{ days}$$

iv. Irrigation efficiency, $I_i = E_a x E_u$

$$= \frac{90x90}{100x100}$$

$$= 81\%$$

Gross irrigation requirement for the irrigation interval

$$IR_g = \frac{ET_{crop}}{Ii} x3days$$

$$= \frac{5.8mm/day}{0.81} x3days = 21.48mm$$

$$\therefore q_d = \frac{21.48mmx5mx5m}{15h}$$

$$IR_g = \frac{ET_{crop}}{Ii} x3days$$

$$= \frac{5.8mm/day}{0.81} x3days = 21.48mm$$

$$\therefore q_d = \frac{21.48mmx5mx5m}{15h}$$

$= 35.8l / h$, assuming duration of irrigation, $I_d = 15h$

$= 36l / h(approx)$

The discharge of 36l/h can be had by using the 6-exit distributor with a discharge of 6l/h per emission point clustered around the plant.

v. The wetted perimeter can be determined by using the Table 3.4 and the following equation

$$\frac{P}{100} = \frac{nS_{ep}S_w}{S_tS_r}$$

Using the discharge rate of distributor 6l/h (average of 4l/h and 8l/h) the width of the wetted strip, S_w, found to bewhich corresponds to S_l value giving P=100% in Table 3.4 for medium soil.

Let spacing between distributor, S_{ep}=1.5m

$$P = \frac{100x6x1.5x1.35}{5x5} = 48.6\%$$

The P value does not match to desired value.

Let specing between distributor, S_{ep}=1.2m

$$P = \frac{100x6x1.2x1.35}{5x5} = 38.88\%$$

The value of P found to be within the desired range.

vi. The layout may be any of the following

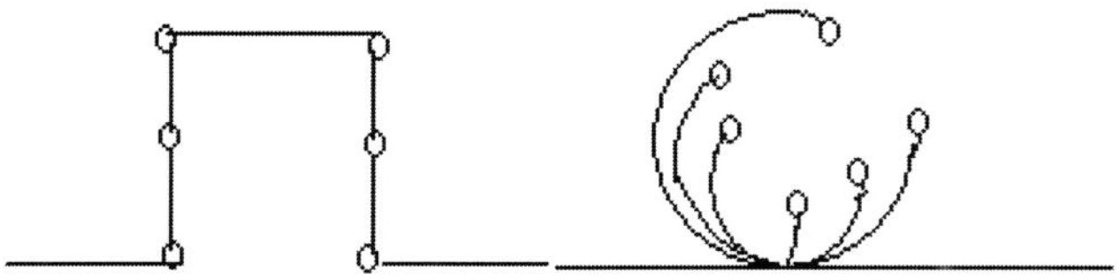

System Capacity

The capacity of the system depends on the type of distributors and the type of distribution methods. Distributions are done either rotational or on demand basis.

Rotational method

In rotational method the entire farm area is divided into number of blocks and the irrigation is done on rotational basis to these blocks. The possible maximum number of blocks in a farm may be determined as

$$N \leq \frac{I_i\, 24\,\text{hours}}{I_d} \tag{3.19}$$

Where, N = number of rotational blocks

I_d = duration of irrigation in each block

It is not expected that the system should operate round the day through the season. There should have some factor of safety to overcome time loss due to breakdown, maintenance, decrease of distributors' discharge, etc. Therefore, for the peak period the maximum operating time hours are taken 20. Thus the capacity of the system,

$$Q_s = \frac{S}{N}\frac{IR_g}{I_d} I_i \tag{3.20}$$

where, Q_s = system capacity

S = total farm area

Example 3.13 Find out the (i) number of rotational units and (ii) system capacity for drip irrigation of a 25ha farm for irrigation interval of 3 days and the duration of irrigation in each irrigation block is 15 hours. Assume gross irrigation requirement as 7mm/day.

i. Number of rotational units,

$$N \leq \frac{I_i \text{x} 24\text{hours}}{I_d}$$

Assuming a 20hr system operation per day,

$$N = \frac{3\text{x}20\text{h}}{15\text{h}} = 4$$

ii. System capacity

$$Q_s = \frac{S}{N}\frac{IR_g}{I_d}I_i$$

$$= \frac{25ha x 7mm/day\ x\ 3days}{4x15h}$$

$$= \frac{5250m^3}{60h}$$

$$= 87.5m^3/h$$

System capacity and week-end consideration

A peak period operation hour has been considered 20 hours. Sometime the system may even works for 22 or even 24 hours to satisfy the field requirements. The allowance of 4 hours may be sufficient for filter clearance or any other maintenance job. However, this is all considered for a week of 7 days. It is desirable to have at least one non-working day in a week. The total requirements of 7 days need to be discharged in 6 days. The time of operation in a day already assumed high which should not be further increased to minimize the requirement of weekend. Therefore, the capacity of the system as calculated on 7 days basis is multiplied by a factor, k depending on the irrigation interval (Table 3.9).

Table 3.9: Values for multiplication factor k

Maximum interval between irrigation in days	Multiplication factor k
2	2
3	1.5
4	4/3
5	5/4
6	6/5
7	7/6
8	7/6
9	7/6
10	7/6

System capacity as per demand basis

In this system irrigation is given as asked by the farmers whenever they want. Therefore, there is a great chance of operating many units at a time. This understandably deserves more system capacity than rotational method where water is applied at some predetermined interval. Clement's suggested the following formula based on statistical probability to find out the capacity of the system (FAO, 1980):

$$x = \frac{1}{r}\left[1 + U\sqrt{\frac{1}{n_1} - \frac{1}{n}}\right] \tag{3.21}$$

Where, x = the ratio Q_d/Q_r in which Q_r is the discharge of the supply line on rotational basis calculated for a continuous irrigation period. Q_d is the discharge of the supply line calculated on as per demand basis.

r = operation factor, daily irrigation time/24 hours

U= parameter depending on the required quality of operation

n_1= number of off-takes simultaneously in operation, $n_1 = \frac{Q_r}{mr}$ in which m is the hydro module.

n = the total number of off-takes.

The daily irrigation time is a factor depends on the habit of the farmers. If the farmers do not like to irrigate in day- time or in night the value of x will reduce. However, the value of r should not go below 0.67. The quality of operation factor, U, reflects the possibility of availability of water in required quantity in any off-takes in the system for which it was designed. For an operational quality of 99 percent the probability of discharging insufficient water is only 1 percent. The operational quality is usually kept between 95 to 99 percent. The operational quality and the corresponding U values are given in Table 3.10. The draw back of Clement's method is that it assumes all the off-takes have the same operational frequency and discharge.

Table 3.10: Values for parameter U

Quality of operation, %	Parameter U
70	0.525
80	0.842
85	1.033
90	1.282
95	1.645
99	2.327
99.5	3.09

Example 3.14 Determine the system capacity as per demand basis using the data in Example 3.12 and the following:

Farm area=420ha

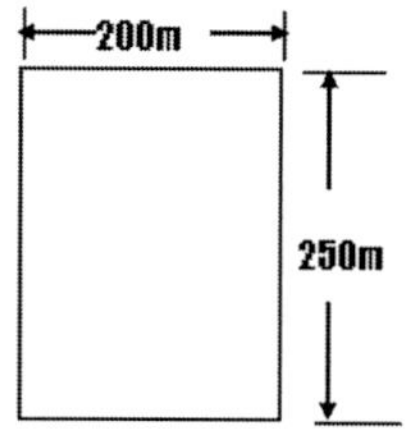

Number of farmers: 20, each of them having 21ha, which for their convenience divide into 4 units of 5ha, 1ha is kept for other purposes.

Layout of the area: Rectangular as shown in figure below.

Hydrant: Each farm unit having one hydrant. The total number of hydrant =20

Laterals: Each small unit of 5ha having 250m/5m spacing = 50 laterals

Distributors: Each lateral is carrying 200m/5m spacing = 40 distributors. Each distributor is discharging 36l/h and lateral discharge is 1440l/h. All 40 laterals are operating at same time.

Hydrant capacity: Each lateral capacity x number of lateral in a sub unit.

$$=1440 \text{ l/h} \times 50$$

$$=72{,}000 \text{ l/h}$$

Hydro module, $= \dfrac{72000}{3600\,\text{sec}} = 20\text{l}/\text{s}$

Specific discharge: Gross irrigation requirement is 7.16mm/day (say 7.2mm). Therefore, the specific discharge is 72000 l/day/ha;

$$= \frac{72000}{3600 \text{x} 24} = 0.83\text{l}/\text{s}/\text{ha}$$

$$Q_r = \frac{400\text{ha} \text{x} 0.83\text{l}/\text{s}/\text{ha}}{3} = 110.67\text{l}/\text{s}$$

The operation factor, $r = \dfrac{15\text{h}}{24\text{h}} = 0.625$

$$n_1 = \frac{Q_r}{m.r} = \frac{110.67\text{l}/\text{s}}{20\text{l}/\text{s} \text{x} 0.625} = 8.85$$

Quality of operation parameter (U): Assuming quality of operation 95%, quality of operation parameter, U=1.645 (Table 3.11).

$$x = \frac{1}{r}\left[1 + U\sqrt{\frac{1}{n_1} + \frac{1}{n}}\right]$$

$$= \frac{1}{0.625}\left[1 + 1.645\sqrt{\frac{1}{8.85} - \frac{1}{20}}\right]$$

$$=1.6\left(1 + 1.645\sqrt{0.113 - 0.05}\right)$$

$$=1.6\left(1 + 1.645 \text{x} 0.25\right)$$

$$1.6\left(1 + 0.41\right)$$

$$=2.256$$

Capacity of the system, $Q_d = Q_r.x$

$$= 110.67 \text{ l/s} \times 2.256$$

$$= 249.67 \text{ l/s}$$

3.0 Questions & Problems

3.1 Define crop water requirement, crop coefficient, net irrigation water requirement, gross irrigation water requirement and peak irrigation requirement.

3.2 How ground coverage is taken in to consideration for water requirement of crop?

3.3 Differentiate the peak irrigation requirement and day-to-day irrigation requirement.

3.4 Discuss the procedure of determination of water requirement of crop under drip irrigation.

3.5 What are the factors govern the wetting pattern of soils in drip system? Why knowing wetting pattern is important? Describe the empirical formulae for determining wetting pattern.

3.6 What is the importance of using multiple emissions? Give the formula relating number of emission points and their spacing.

3.7 Discuss the irrigation at fixed deficit and fixed interval.

3.8 Explain the impact of rate of evapotranspiration on frequency of irrigation.

3.9 Explain the influence of week-end on system capacity.

3.10 Describe the method of determining the system capacity in rotational method and as per demand basis of drip irrigation.

3.11 Find the scope and limitation of guideline provided by Keller and Karmelli (1974) in designing the lateral spacing and percentage of area of wetting.

3.12 The following are the pan evaporation (E_{pan}), pan factor (k_p) and crop coefficient (k_c) of an arbitrary location and crop.

	Jan	**Feb**	**Mar**	**April**	**May**	**June**	**July**	**Aug**	**Sep**	**Oct**	**Nov**	**Dec**
E_{pan} mm/day	2.4	2.5	3.9	4.1	4.9	5.0	4.7	4.6	4.5	3.9	2.6	2.3
k_p	0.75			0.8				0.75				
k_c	0.5	0.6	0.65	0.7	0.75	0.8	0.85	0.85	0.9	0.85	0.75	0.65

Determine the water requirement of crop (ET_{crop}) at different months. Assume 40% ground coverage from January to April, 70% from May to August and 80% for the rest of the months.

3.13 A crop was grown at 5m x 5m spacing. The pan evaporation in a day was recorded 5mm. If the K_p, K_c and K_r were 0.8, 0.7 and 0.6 respectively, what was the crop water requirement in l//plant for that day?

Ans. 42

3.14 In a research work for the determination of water requirement of a crop, the yields of the crop were 42,550, 45,650 and 30,345 kg/ha for the irrigation treatments at 80, 60 and 40 percent of pan evaporation respectively. During the experimentation the cumulative pan evaporation was recorded 480mm.Compare the water use efficiency of different irrigation treatments.

Ans. 1108, 1585 & 1580 kg/ha/cm respectively

3.15 Determine the probable width and depth of wetting when drip irrigation is done from a point source at the rate of 4 l/h and 5 l/plant/day. The saturated hydraulic conductivity of the soil is 1.1 m/day.

Ans. Depth = 0.245m, width = 0.388m

3.16 Ten liter water has been applied from point source at a rate of 1.5 l/h in clay soil. Estimate the possible width and depth of wetting of soil.

Ans. width = 0.94m, depth = 0.73m

3.17 Design the spacing of inner rows of laterals and find the percentage of wetting of a field for a crop spaced 5mx5m in fine textured soil assuming the discharge rate of distributors as 4 l/h.

Ans. Inner spacing = 1.5m, Percentage of wetting = 62.2

3.18. The tree plants are cultivated at 6.0mx6.0m spacing in coarse textured soil. The inner space of paired laterals through the plant rows is 1.2m. Find the percentages of area of wetting if the distributors used are of capacity 4 l/hr.

Ans. 26.84%

3.19. The tree crops spaced at 6.5m x 6.5m in medium textured soil using 4 numbers of distributors in each plant and each of capacity 6 l/hr. The spacing of distributors is 2.5m. Determine the possible percentage of area of wetting.

Ans. 31.95%

3.20. Determine the PIR from the following data:

FC = 20%, PWP = 7%, BD = 1.4gm/cm^3, root zone depth = 1.2m, maximum allowable soil moisture depletion = 30%, design objective to wet the potential root zone = 45%, irrigation interval = 4 days.

Ans: 7.38 mm/day

3.21. Find the number of units of a rotational drip irrigation system where the irrigation interval is 3 days and each distributor is operated for 10 hours.

Ans: 6

3.22. What is the discharge per distributor or group of distributor if gross irrigation requirement is 27.0mm, spacing of plants 5m x 5m and the distributors are operated for 15 hours in each irrigation.

Ans. 45 l/h

3.23. Determine the system capacity in drip irrigation from the following data:

Area of farm: 120ha

No. of units = 4

Net irrigation requirement = 6.2mm/day

Irrigation efficiency = 90%

Irrigation interval = 4 days

Assume any other data if necessary.

Ans: 153.09 l/s

3.24. A drip irrigation system for 100ha area is designed for maximum hours of operation 20 in a day during peak period and the distributors are operated 16 hours in each irrigation. What is the suitable number of irrigation units for the area? What is the system capacity if the irrigation interval is 4 day?

Ans: 5 units, 90.29 l/s

3.25. Write True or False of the following:

1. Reference crop evapotranspiration is always more than evapotranspiration.
2. Crop coefficient is usually more in early stage of plant growth.
3. Oasis effect to ground coverage factor (k_r) is required in early days of plant growth.
4. In any given situation the irrigation water requirement and crop water requirement may be same.

5. All that rainfall occurs can be subtracted from crop water requirement to calculate net irrigation requirement.
6. The drip irrigation system is designed on the basis of peak irrigation requirement.
7. The pan factor fluctuates around 0.8.
8. The amount of water to be applied in any irrigation is calculated by using the modified values of k_c, k_r and E_{pan}.
9. Light soil is susceptible to deep penetration of water.
10. Point source of application of water is better to application of water in tree plants.

Ans. 1. False 2. False 3. True 4. True 5. False 6. True 7. True 8. True 9. True 10. False.

3.26. Select the appropriate answer from the following.

1. In any irrigation system, the primary objective is to know the
   a) Water requirement of crop b) Soil type
   c) Source of water d) Season of the year
2. The beneficial loss among the following is
   a) evaporation b) Seepage
   c) Transpiration d) Runoff
3. Lowest among the following is
   a) Irrigation water requirement
   b) Net irrigation requirement
   c) Gross irrigation requirement
   d) Crop water requirement
4. The irrigation system capacity is selected considering the
   a) Absolute peak pan evaporation
   b) Ninety percent of peak pan evaporation
   c) Average of peak pan evaporation of the week
   d) Average of the month of peak pan evaporation
5. At any rate of drip water application the maximum horizontal spreading is expected in
   a) Coarse soil b) Medium soil
   c) Light soil d) Heavy soil

6. Application of water by single distributor may not cause suffer to
   a) Shallow rooted crops b) Medium rooted crops
   c) Deep rooted crops d) Tree plants
7. Rate of water application is usually low in
   a) Light soil b) Medium soil
   c) Heavy soil d) Deep soil
8. Week-end effect depends on
   a) Net irrigation requirement b) Type of soil
   c) Irrigation interval d) Ground cover
9. Duration of irrigation in each block influence to select the
   a) Number of rotational block
   b) Duration of system operation in each day
   c) Irrigation interval
   d) Gross irrigation requirement
10. Hydro module is usually expressed in
   a) m^3/day b) m^3/s
   c) l/h d) l/s

Ans. 1. (a) 2. (c) 3. (b) 4. (d) 5. (d) 5. (a) 6. (a) 7. (c) 9. (a) 10. (d)

References

Biswas, R.K., Rana, S.K. and Mallick, S.(1999). Study on the performance of drip irrigation in papaya cultivation in New Alluvial agro-climatic zone of West Bengal. J.Ann.agric. Vol.20(1);116-117.

FAO (1980). Vermeiren, I and Jobling, G.A. Irrigation and Drainage Paper 36, FAO. Rome.

Karmeli, D. and Keller, J. (1975). Trickle Irrigation Design. Rain Bird Sprinkler Manufacturing Corp., Glendora, California. pp.133.

Schwartzman, M and B. Zur (1985). Emitter spacing and geometry of wetted soil volume. J.Irri. Drainage Engr, ASCE, 112 (3):242-253.

4

Design of Pipe Network

The pipe network in a drip system consists of a mainline, sub mains, laterals, and distributors. The layout of the pipes depends on the physical factors like shape and size of the area, topography and any other obstacles in the field. In any field there may be the scope of different layout of the pipes. The most economic one should be selected. In general, contour line is followed in laying the laterals in sloping land, sub main length usually be shorten in uphill and equally spaced laterals on either side of a sub main in a flat land provides good layout.

Hydraulic Formulae /Head Losses in Pipes

There are numerous equations for solving the head loss in pipes. The most commonly used are the Darcy-Weisbach, Hazen-William and Scobey's.

Darcy-Weisbach equation

$$h_l = f\frac{l}{d}\frac{v^2}{2g} \tag{4.1}$$

Where, $f = \frac{0.3164}{R_e^{0.22}}$ (for turbulent flow)

or, $f = \frac{64}{R_e}$ (for laminar flow)

Hazen-William equation

$$h_l = \frac{3.022 l v^{1.852}}{C^{1.852} d^{1.167}} \tag{4.2}$$

Where, h_l = head loss of the pipe of length l, m

v = velocity of flow, cm/s

d = diameter of pipe, m

Scobey equation

$$h_l = \frac{K_s l v^{1.9}}{1000 d^{1.1}} \tag{4.3}$$

In the above equations,

h_l = the friction loss of l length of pipe

v = the mean velocity

d = the diameter of the pipe

C,f,Ks = constants.

The Eq. 4.1 to 4.3 may be generalized as follow:

$$h_l = K_1 \frac{lv^m}{d^n} = K \frac{lq^m}{d^{2m+n}} \tag{4.4}$$

Where, q = discharge in the pipe

m, n = exponents depends on the formula used

K = constant depends on the formula used

When Scobey's equation is used, m = 1.9 and n = 1.1.

William-Hazen equation:

The William-Hazen equation for smooth pipe (using C=150 which are normally used for plastic pipes) is as follows:

$$h_l = 1.135 \times 10^6 \frac{q^{1.852}}{d^{4.871}}.l \text{ or } h_l = 15.27 \frac{q^{1.852}}{D^{4.871}}.l \tag{4.5}$$

in which h_l and l in m, q is the total discharge in l/s, d is the internal diameter in mm and D in cm.

The Eq. 4.5 converts to

$$h_l = 3.98 \times 10^5 \frac{q^{1.852}}{d^{4.871}}.l \tag{4.6}$$

for lateral and sub-main where flow decreasing to zero at the end of the pipe. The Eq. 4.6 is applicable to lines with more than twenty outlets.

When C is given, the total energy drop can be calculated by using following formula

$$h_l = \left(\frac{Q}{C}\right)^{1.852} D^{-4.871} LF \tag{4.7}$$

Where, h_l is the head loss in pipe, m

K is constant, 1.21x1010

Q is the flow rate in the pipe, l/s

C is the coefficient of friction for continuous section

D is the diameter of pipe (inside), mm

F is the outlet factor

L is the length of pipe, m

The Table 4.1 gives the C value of Hazen-William formula for different types of pipes. The friction losses can also be determined from the equations in Table 4.2 and the Fig. 4.1-4.3. These values are obtained following Hazen-William equation for C=150 which is the normally used value in plastic pipes.

Table 4.1: Hazen and Williams coefficient for flow in pipes

Description of pipe	C Value
Polyvinyl chloride pipe	155
Extremely smooth and straight	140
Very smooth	130
Smooth wood and wood stave	120
New riveted steel	110
Vitrified	110
Old riveted steel	100
Old cast iron	100
Old pipes at bad condition	60 to 90
Small pipes with badly tuberculated	40 to 50

Courtesy: Sivanappan, R.K.(1984)

Table 4.2: Equations to calculate the friction losses in PVC pipes (C=150) following Hazen-William

ID of pipe, mm	Head losses (AH), m/lOOm; (q, 1/s)
8.5	3353.24 lg^{1852}
8.8	2831. 9741,?$^{1\ 852}$
9.0	2538.33g^{L852}
9.4	2053.8,?^{1852}
12.4	493.2192g^{1852}
15.2	197.6534g^{1852}
16.4	136.5084^{1852}
19	66.6585?^{1852}
20.8	51.9215^^{1852}
24.2	20.51 *63q*$^{l\ 852}$
26.8	12.4801g^{1852}
46	0.8982$_{?}^{1852}$

56	$0.3445?^{'852}$
71	$0.1\ 084V^{852}$
84	$0.0478g^{1852}$

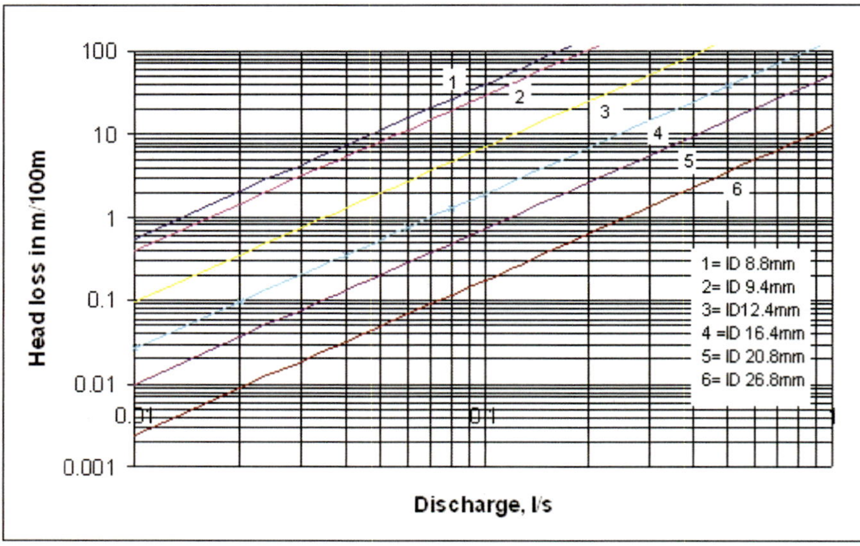

Figure 4.1: Head losses in PVC pipes (C=150) following Hazen-William

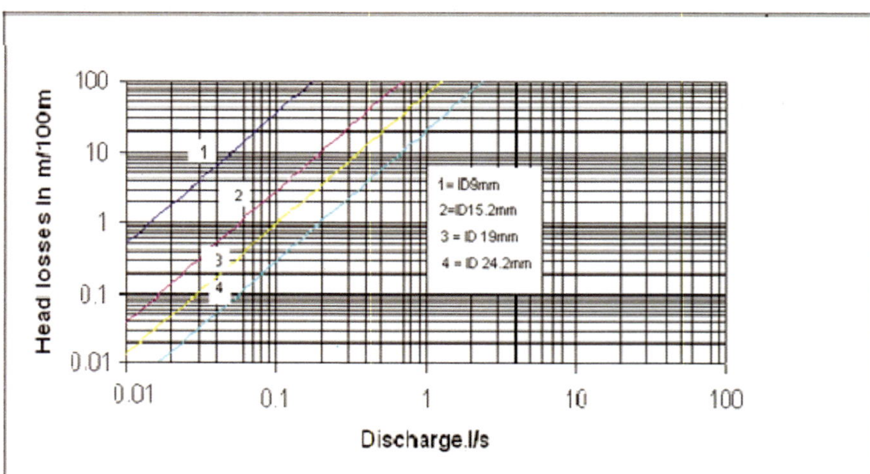

Figure 4.2: Head losses in PVC pipe (C=150) following Hazen-William

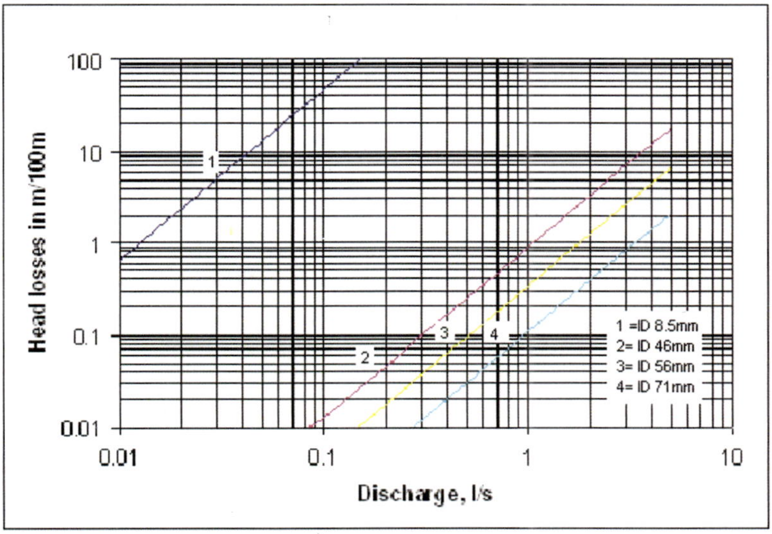

Figure 4.3: Head losses in PVC pipe (C=150) following Hazen-William

Lateral Design

Design of the lateral depends on the pressure distribution in it and the characteristics of the distributors. Uniformity of irrigation can be achieved by adjusting the size of the distributors or length of the micro tubes. The pipe size of the laterals can be determined on the following ways:

Analytical method

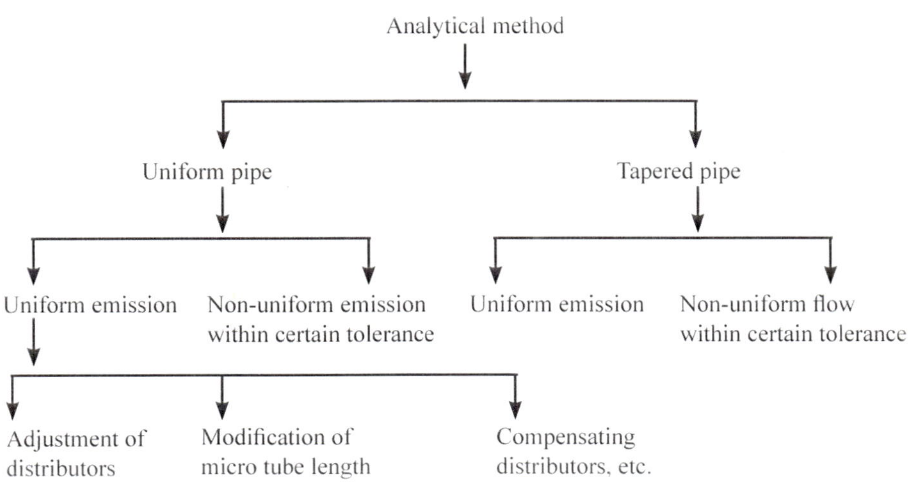

Graphical method

The diameter of a lateral is selected on the basis of difference of discharges of the distributors should not exceed 10 percent.

Analytical method

a. Head losses in a pipe with decreasing discharge

i. Uniform pipe flow

The drip laterals consists of multiple flow outlets. The friction losses through these pipes can be determined by the formula proposed by Christiansen (1942) as below:

$$\Delta H = F\left[\frac{klQ^m}{d^{2m+n}}\right] = F\left[\frac{k_1 l v^m}{d^n}\right] \tag{4.8}$$

Where, F is a reduction coefficient whose value depends on the number of outlets on lateral. Friction loss of a lateral is first determined by using the Eq. 4.1 to 4.3 as if there is only one outlet in a lateral. The friction loss is multiplied by the coefficient F to determine the friction losses of laterals with multiple outlets. The Table 4.3 gives the value of F for the formulae as described in Eq. 4.1-4.3.

Table 4.3: Values of coefficient F

Number of outlets	**Hazen-Williams m = 1.85**	**Scobey m = 1.90**	**Darcy-Weisbach m = 2.0**
1	1	1	1
2	0.639	0.634	0.625
3	0.535	0.528	0.518
4	0.486	0.480	0.469
5	0.457	0.451	0.440
6	0.435	0.433	0.421
7	0.425	0.419	0.408
8	0.415	0.410	0.398
9	0.409	0.402	0.391
10	0.402	0.396	0.385
11	0.397	0.392	0.380
12	0.394	0.388	0.376
13	0.391	0.384	0.373
14	0.387	0.381	0.370
15	0.384	0.379	0.367
16	0.382	0.77	0.365

17	0.380	0.375	0.363
18	0.379	0.373	0.361
19	0.377	0.372	0.360
20	0.376	0.370	0.359
22	0.374	0.368	0.357
24	0.372	0.366	0.355
26	0.370	0.64	0.353
28	0.369	0.363	0.351
30	0.368	0.362	0.350
35	0.365	0.359	0.347
40	0.364	0.357	0.345
50	0.361	0.355	0.343
100	0.356	0.350	0.338
∞	0.351	0.345	0.333

Courtesy: FAO (1980)

Considering the F value the following general equation was proposed by Wu and Gitlin (1975) as below which gives the approximate calculation of total head losses in a lateral.

$$h_l = 5.35\left(\frac{q^{1.852}}{d^{4.871}}\right)l \tag{4.9}$$

Where, h_l = total energy drop by friction at the end of the lateral or submain, m

q = total discharge at the inlet of lateral or submain, l/s

d = inside diameter of lateral or submain, cm

l = length of the lateral or submain,m

ii. Tapered pipe lateral

The smaller pipe size can be used to down streams of the long laterals as the flow decreases. This provides saving in cost of pipes. The total head losses in such a lateral may be determined by

$$\Delta H_t = \Delta H_{d1l} - \Delta Hd_1l_2 + \Delta H_{d2l2} \tag{4.10}$$

Where, ΔH_t = total head loss in tapered pipe

$\Delta H_d 1_l$ = the head loss for the total length considering the larger diameter

$\Delta H_{d2} l_l$ the head loss for the last portion considering the larger diameter

the head loss for the last portion considering the smaller diameter.

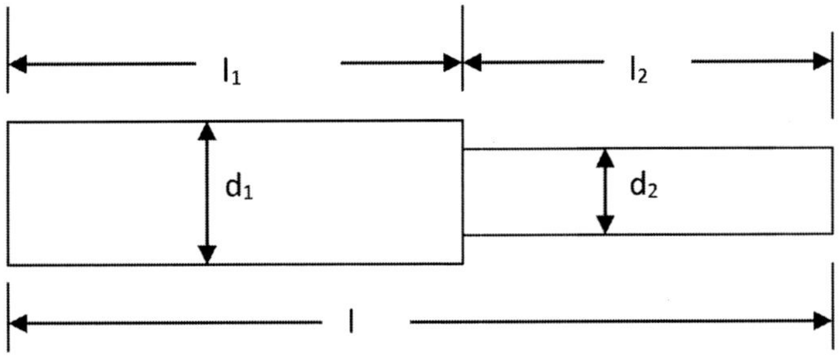

Figure 4.4: Tapered lateral pipe

b. Effects of temperature and local head loss

Hazen-William formula is widely used in irrigation and water-works design and is most applicable in pipes of diameter 5cm and larger, and velocities less than 3.1m/s. The advantage of using this formula is that the coefficient C does not involve Reynolds number and all problems have direct solution. However, this is also the disadvantage, since this avoids the effects of temperature and thereby the viscosity. Sometime this causes serious error to design. Therefore, when Hazen-William formula is being used the design may be verified by Darcy-Weisbach formula and multiplied with a coefficient reflecting the temperature effect.

The type of distributors used along the lateral has got the influence on total head losses of the lateral. Different type of distributors offers different local resistance to flow. Considerable resistance is obtained for in-line or bayonet type of distributors. The following formula determines the local head loss in a lateral with in-line distributors.

$$\Delta h = \frac{\sigma q^2}{4650 r^4} \tag{4.11}$$

Where, q = the nominal discharge of the distributors, m^3/s

r = the inside diameter (ID) of the distributors, m

σ = a coefficient the value of which depends on the number of distributors on the lateral and is defined as

$$\sigma = 2e^3 + 2.73e^2 - 0.70e \tag{4.12}$$

and, e=number of distributors on the line.

Biswas *et al.* (2005) described the local head losses due to protrusion of distributors or sub-laterals in to laterals as barb losses. They studied the pressure heads in the laterals with and without emitters/sub-laterals, the discharges, velocity of flow and friction head losses and thereby the barb losses. Barb loss due to each emitter was found 0.038m, which was equivalent to friction loss in 21.77 cm length of lateral pipe. The barb loss due to 4mm sub-lateral was 0.046m. Reddy *et al.* (2001) reported comparable result of 16.804cm equivalent length of lateral as barb loss for each emitter.

c. Pressure distribution along a lateral

The total specific energy at any section of a drip line can be expressed by the energy equation

$$\bar{H} = z + H + \frac{v^2}{2g} \tag{4.13}$$

Where, $\bar{H}$ = total energy, m

H = pressure head, m

z = potential head or elevation, m

v = velocity of flow, m/s

g = acceleration due to gravity, m/s2.

The energy gradient line in drip is not like a straight line but an exponential type with respect to the length. The shape of the energy gradient line may be expressed by the following equation

$$Ri = 1 - (1 - i)^{m+1} \tag{4.14}$$

Where, $R_i = \frac{\Delta H_i}{\Delta H}$ and is called the energy drop ratio

ΔH_i total pressure drop at a length ratio $i\left(i = \frac{l}{L}\right)$

ΔH the total energy drop at the end of the line

L = total length of the line

l = given length measured from the head end of the line

m = exponent of the flow rate in friction equation.

The dimensionless energy gradient lines are the followings for different friction equations

$$R_i = 1 - (1 - i)^3 \text{ (following Darcy-Weisbach)} \tag{4.15}$$

$$Ri = 1 - (1 - i)^{2.852} \text{ (following Hazen-William)} \tag{4.16}$$

$$Ri = 1 - (1 - i)^{2.9} \text{ (following Scoby)} \tag{4.17}$$

The change of energy in a drip line can be had by differentiating Eq. 5.76 with respect to length.

$$\frac{d\bar{H}}{dL} = \frac{dz}{dL} + \frac{dH}{dL} + \frac{d}{dL}\left(\frac{v^2}{2g}\right) \tag{4.18}$$

Since the discharge of the distributors along the drip line is low, the change of velocity head with respect to length is negligible. Therefore, Eq. 5.81 becomes

$$\frac{d\bar{H}}{dL}=\frac{dz}{dL}+\frac{dH}{dL} \tag{4.19}$$

Where, $\frac{dH}{dL}=-S_f$ (4.20)

The negative sign indicates the friction loss with respect to length. The ratio $\frac{dz}{dL}$ represents the slope of the line, as

$$\frac{dz}{dL}=\pm S_o \tag{4.21}$$

Therefore $\frac{d\bar{H}}{dL}=-S_f+S_o$ (4.22)

When the lateral is laid to down slope.

$$\frac{d\bar{H}}{dL}=-S_f-S_o \tag{4.23}$$

When the lateral is laid to up slope.

The pressure variation along the lateral can be expressed as below if the input pressure is given,

$$H_i=H-R_i\Delta H\pm R_i^{'}\Delta H' \tag{4.24}$$

Where, H_i = hydrostatic head at a given length ratio i,

H = input pressure

ΔH_lfrictional loss of head at a given length ratio i,

$\Delta H'_l$the energy gain (down slope) or loss (up slope) at a given length ratio i.

The Eq. 4.24 can be expressed by using energy drop ratio R_i and the energy gain or loss ratio R'_i,

$$H_i=H-R_i\Delta H\pm R_i^{'}\Delta H' \tag{4.25}$$

Where, ΔH = total energy drop by friction

ΔH = total energy gain or loss by slope

$$R_i=\frac{\Delta H_i}{\Delta H}$$

$$R_i^{'}=\frac{\Delta H_i^{'}}{\Delta H^{'}} \tag{4.26}$$

For uniform slopes, R'_i is the same as the length ratio i. Therefore,

$$H_i = H - Ri\Delta H + R'_i \Delta H' \tag{4.27}$$

For non-uniform terrain with $S_1, S_2, S_3, \ldots S_j \ldots S_n$ slopes of the sections,

$$H_i = H - R_i \Delta H + \frac{L}{n}\sum_{1}^{n} S_j \ldots \left(S_j = 1, 2, 3, \ldots n\right) \tag{4.28}$$

Where, S_j = slope of the j section along the line using '+' sign for down slope (energy gain) and '-' sign for up slope (energy loss).

Example 4.1 Determine the pressure and emitter flow variation of a 15mm (ID) lateral line of 100m length. The emitters are 0.80m spaced along the lateral and emitter flow is 4l/h at an operating pressure of 10m at head end. Lateral line is 1.5% down slope, the x-value in the q-H relationship of the emitters to be 0.5.

Solution: Lateral length line, L=100m

Lateral line size, D=15mm (ID)

Total lateral flow,

$$q = \frac{100m}{0.8m}.4l/h$$

$$= 500l/h$$

$$= 0.139l/s$$

The total pressure drop (friction loss) using Eq. 4.6,

$$\Delta H = 3.98 \times 10^5 \times \frac{q^{1.852}}{d^{4.871}}..l$$

$$= 3.98 \times 10.^5 \times \frac{(0.139)^{1.852}}{(15)^{4.871}}.100$$

$$= 3.98 \times 10^5 . \frac{2.59 \times 10^{-2}}{5.35 \times 10^5}.100$$

$$= 1.923m$$

At $i = 0.1$, $R_i = 1 - (1 - i)^{2.852}$ [following Eq. 4.16]

$= 1 - 0.9^{2.852}$

$= 0.26$

$$\text{and}, H_i = H - R_i \Delta H + \frac{L}{n}\sum_{1}^{n} S_j$$

or, $H_{0.1} = 10 - 0.26 \text{ x } 1923 + 10 \text{ x } \frac{1.5}{100}$

$= 9.65m$

Similarly, at i = 0.2, H = 10 - 0.471 x 1.923 + 0.45 = 9.39*m*

i = 0.3, H = 10-0.638x1.923+ 0.45 = 9.22*m*

i = 0.4, H = 10-0.768x1.923+ 0.60 = 9.12*m*

i = 0.5, H = 10 - 0.861 x 12.923 + 0.75 = 9.09*m*

i = 0.6, H = 10 - 0.927 x 12.923 + 0.90 = 9.11*m*

i = 0.7, H = 10-0.968x21.923+ 1.05 = 9.18*m*

i = 0.8, H -10 - 0.99 x 1.923 +1.20 = 9.29*m*

i = 0.9, H = 10-0.999x1.923+ 1.35 = 9.42*m*

i = 1.0, H = 10-1.0x1.923+ 1.50 = 9.57*m*

Pressure variation, $H_{var} = \frac{H_{max} - H^{min}}{H_{max}} = \frac{10 - 9.09}{10} = 0.091$

Emitter flow variation, qvar = 1 - (1 - Hvar)x

$= 1 - (1 - H_{var})^x$

$= 1 - (1 - 0.091)^{0.5}$

$= 1 - 0.95$

$= 0.0466$

Keller and Karmeli method

The average distributor outlet is defined as the outlet, which has the average discharge rate, and the pressure head, which causes this average discharge, is called the average head. Keller and Karmeli (1974) studied the wide range of distributor exponents and pressure losses and found that the average pressure occurs at the length ratio i = 0.39 following FAO (1980) (Fig. 4.5). They also found that approximately 77 percent of the total head loss occurs between i=0.0 and i=0.39, and the rest 23 percent occurs between i=0.39 and i=1.00. Therefore, for a flat terrain average head loss can be determined by

$$\bar{H} = H(i=0) - 0.77\Delta H \tag{4.29}$$

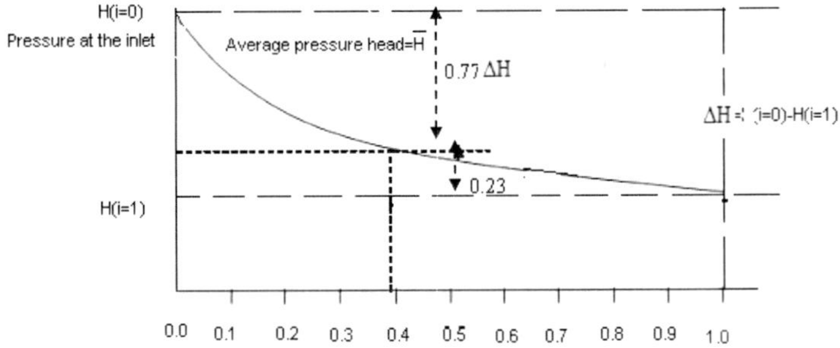

Figure 4.5: Pressure distribution along a lateral

If the discharge-pressure relationship ($q = K_d H^x$)of the distributors are available by the manufacturers then the q and q_{min} can be determined since

$\bar{H}$ and H_{min} are known. This information will enable to calculate the emission uniformity (E_u).

The following equations give very good fit to head loss curve for a constant diameter lateral pipe in flat terrain,

$$H(i)=H(i=1)+E(i)\,\Delta H \tag{4.30}$$

Where, is the fraction of the total head remains to be lost to the end of the line. At The values of for various values of i are given in Fig. 4.6 & 4.7. These were obtained by resolving the equations for *i*.

$$E(i)=\exp\left[-4.38815.\frac{(i)^{1.19088}}{(1-i)^{0.05555}}\right] \tag{4.31}$$

Example 4.2 Determine the uniformity of emission and the pressure curve on flat, sloping and undulated terrain with the following:

Lateral length = 200m

Distributor discharge = 4l/h

Distributor characteristics, $q = K_d H_x = 0.65H^{0.8}$

Average pressure=10m

Manufacturers' coefficient of variation, CV= 4%

PVC pipes available in the market = ID (mm)-9.4, 12.8, 16.4, 20.8.

Spacing of distributors = at every 5m.

Solution

For flat terrain: Number of distributors on the line $= \frac{200\text{m}}{5\text{m}} = 40$

Lateral flow rate = 40x4l/h =160 l/h

= 0.044 l/s

The total head loss in the lateral is tabulated below (Table 4.4) as determined by using Fig. 4.1-4.3 and the appropriate value of F (Table 4.3) assuming no local loss and temperature at 200C.

Table 4.4: ΔH values for different ID

ID (mm)	Head loss m/100m full flow	Total ΔH full flow	F	$\Delta H(m)$	$\Delta H(m)$ following Eq. 4.6	Remarks
9.4	8.7	17.4	0.364	6.33	4.45	Pipe selected should not exceed the tolerance limit of 10% of average pressure
12.8	1.9	3.8	0.364	1.38	0.99	
16.4	0.5	1.0	0.364	0.36	0.30	
20.8	0.14	0.28	0.364	0.10	0.09	

From the above calculation following the Fig. 4.1-4.3 it appears that among the available diameter pipes, the minimum diameter pipe, which satisfies the tolerance limit, is 16.4. However, head loss calculated following Eq. 4.6 marginally approves the 12.8mm diameter pipe. Therefore, 16.4mm diameter pipe is preferred.

$H\,(i{=}1) = 10m - 0.23\Delta H = 10 - 0.23 \text{ x } 0.364 = 9.92m$

$H\,(i{=}1) = 10m - 0.77\Delta H = 10 + 0.28 = 10.28m$

$q^{\min} = 0.65(9.92)^{0.8} = 0.65 \text{ x } 6.31 = 4.10l\,/\,h$

$q = 0.65(10)^{0.8} = 0.65 \text{ x } 6.31 = 4.10l\,/\,h$

Manufacturer's discharge ratio, Mr = 1 - 1.27CV

$$=1 - 1.27 \text{ x } 0.04$$

$$= 1 - 0.05 = 0.95$$

$$E_u = 100\frac{q_{min}}{\bar{q}}.M_r f(e), f(e)=1$$

$$=100\text{x}\frac{4.07}{4.10}\text{x}0.95$$

$$=94.3\%$$

The pressure distribution curve is determined by using the Eq. 4.31. The values obtained are given in Table 4.5 and represented by curve on Fig. 4.6 & 4.7.

Table 4.5: Pressure variation (H(i)) along the lateral

i	E(i)	E(i)ΔH	H(i), meter
0	1	0.364	10.28
0.025	0.99	0.363	10.27
0.1	0.95	0.345	10.26
0.2	0.81	0.296	10.22
0.3	0.64	0.232	10.15
0.4	0.456	0.166	10.09
0.5	0.297	0.108	10.03
0.6	0.175	0.063	9.98
0.7	0.093	0.034	9.95
0.8	0.044	0.016	9.94
0.9	0.017	0.006	9.93
1.0	0	0	9.92

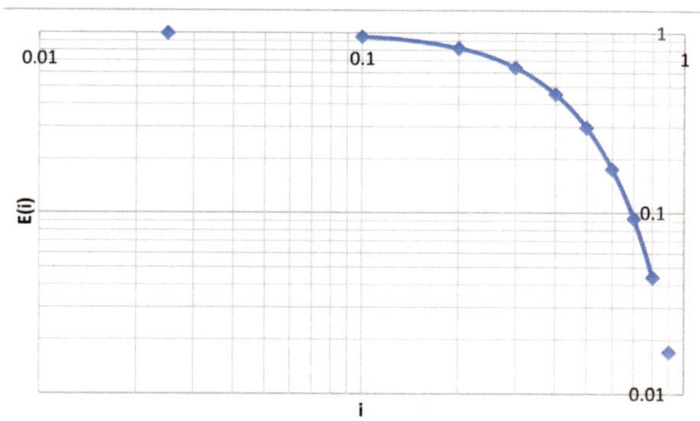

Figure 4.6: Values if E(i) against i in log-log paper

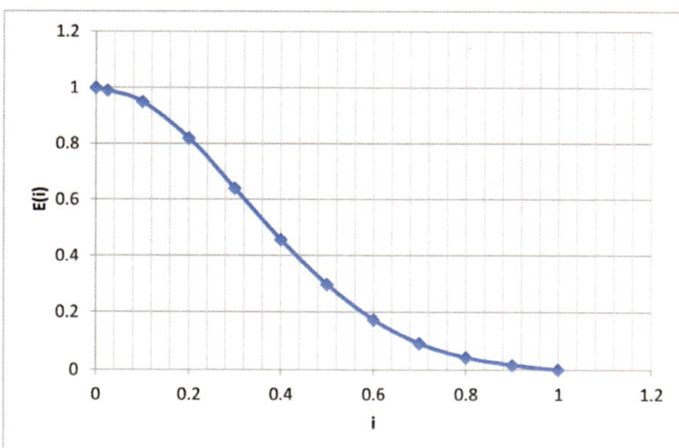

Figure 4.7: Values if E(i) against i in plain paper

Let us assume a uniform slope of 0.1% for sloping terrain and 0.1% up to 100m and 0.1% down after 100m for undulating terrain. By studying the gain or loss in elevation the pressure head at different length ratio (i) and the corresponding discharges are tabulated in Table 4.6 for sloping and undulated terrain.

Table 4.6: Pressure variation (H(i)) for sloping and undulated terrain

Length ratio	Sloping terrain		Undulated terrain	
	Head, m	**Discharge, l/h**	**Head, m**	**Discharge, l/h**
0	10.28	4.19	10.28	4.19
0.1	10.28	4.19	10.24	4.18
0.2	10.26	4.19	10.18	4.16
0.3	10.21	4.17	10.09	4.13
0.4	10.17	4.16	10.01	4.10
0.5	10.13	4.14	9.93	4.08

0.6	10.10	4.13	9.90	4.07
0.7	10.09	4.13	9.84	4.07
0.8	10.10	4.13	9.85	4.06
0.9	10.11	4.14	9.91	4.07
1.0	10.12	4.15	9.92	4.07
		$\sum = 45.72$ $\bar{q} = 4.16$		$\sum = 45.15$ $\bar{q} = 4.11$

Considering the discharges for only the length ratio (i) of 0, 0.1, 0.2,….,1.0, from Table 4.6 the average discharges are 4.16 and 4.11 l/h for sloping and undulated terrain respectively.

Therefore,

The pressure distribution at different lengths for flat, undulated and sloping terrains are shown in Fig. 4.8.

Distance, m	Sloping terrain Head, m	Undulated terrain Head, m	Flat terrain Head m
0	10.28	10.28	10.28
20	10.28	10.24	10.26
40	10.26	10.18	10.22
60	10.21	10.09	10.15
80	10.17	10.01	10.09
100	10.13	9.93	10.03
120	10.1	9.9	9.98
140	10.09	9.84	9.95
160	10.1	9.85	9.94
180	10.11	9.91	9.93
200	10.12	9.92	9.92

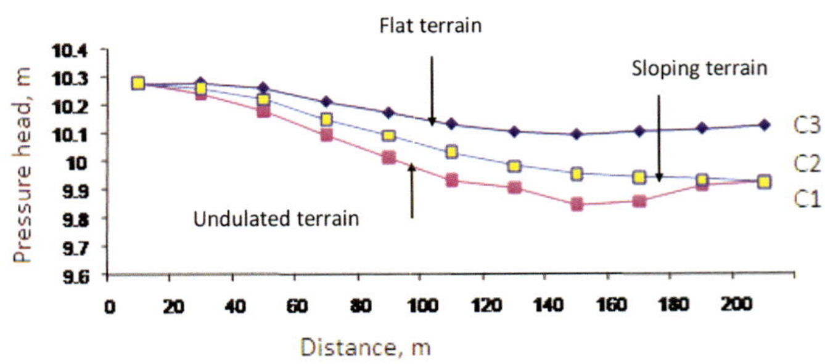

Figure 4.8: Pressure distribution curves for flat terrain C1, sloping terrain C2 and undulated terrain C3

Graphical Design Method

The graphical design method is faster than the methods described earlier. It is therefore more useful for designing large number of laterals. This method uses a set of curves called "polyplot" the development and use of which is described by Herbert (1971). The polyplot technique is based on the following assumptions and limitations:

i. Specific Discharge Rate, SDR, characterizes the flow in the lateral.

$$\text{SDR} = \frac{\text{discharge per distributor or group of distributor}}{\text{Distance between distributor or group of distributors}}$$

$$\text{or, SDR} = \frac{\text{total discharge}}{\text{total length of flow}}$$

The SDR technique is applied only to pipelines having flow decreasing to zero at the closed end.

ii. The spacing between the outlets is 6m. If the outlets are more than 6m, friction loss will be little higher and the friction curve will be slightly above the friction curve shown. Similarly, when spacing is less than 6m, the real curve will be slightly below. However, for most practical purposes these curves are accurate enough.

iii. Friction losses in pipes are calculated based on Darcy-Weisbach equation in which friction f is taken as:

$$f = \frac{64}{R_e} \text{ for laminar flow}$$

and $f = 0.0056 + 0.50R_e^{-0.32}$ for turbulent flow.

The "change-over point' from laminar to turbulent is not stated. However, it may be assumed to be R_e =2000. The Table 4.7 represents the limits of laminar flow for various discharges.

iv. Values of f are established empirically for smooth pipes. Since the f values are established for smooth pipes, the curves are established for smoothly pipes. The curves will be applicable to other commercial pipes, because even for steel and asbestos pipes the friction factor does not deviate much to a quite high Reynolds number.

v. The method does not take into account the possible fluctuation due to lack of homogeneity in the manufacture of distributors. The coefficient of variation due to manufacture is equally important to friction head loss in the system. There is some risk to implement the design unless the emission uniformity is verified.

Table 4.7: Approximate limits of laminar flow for various pipe sizes and discharges

Internal diameter of pipe, mm	Flow at which Re
0.51	2.4 lph
0.89	4.6
1.27	6.7
6.35	0.6 lpm
6.35	0.9
12.70	1.2
15.88	1.5
19.05	1.8
24.40	2.4
31.75	3.1
38.10	3.7
50.80	4.8

Source: Irrigation and Drainage Paper 36, FAO (1980)

The steps required to use the "polyplot" techniques in designing the laterals are stated below with reference to Fig. 4.9.

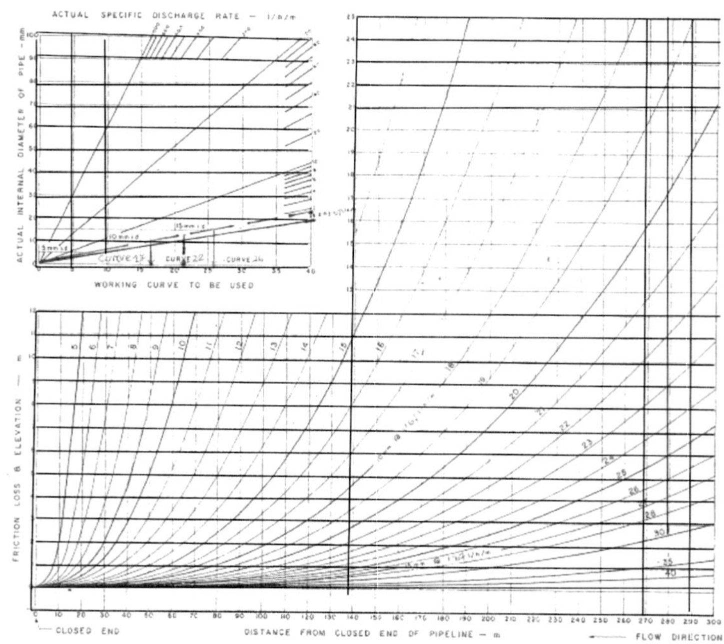

Figure 4.9: Graphical design of pipelines with flow decreasing uniformly to zero

Source: Irrigation & Drainage Paper 36, FAO(1980) (p.63)

Case I. Fixed outlet discharge with tapered pipes.

Step 1. Calculate the SDR and make this SDR line on the small graph in the upper left corner.

Step 2. Select the various pipe sizes (internal diameter) expected to be used on laterals.

Step 3. Mark the points of intersections of SDR lines to lines of diameter of pipes.

Step 4. Determine the Working Curve denomination towards down from the points of intersection for various diameters.

Step 5. Marks Working Curves so determined on the main chart. These are then the actual friction curves for these particular pipe diameters and SDR.

Step 6. Select the nominal design pressure at which the lateral is to be operated.

Step 7. Place a sheet of tracing paper over the chart and draw at same scales ordinate and abscissa axis of the underlying friction curves.

Step 8. Origin of axis is the closed end at the farthest point of the lateral. From the origin draw the original profile of the terrain over which the lateral is to run.

Step 9. Draw the lines parallel to the profile that represents the design tolerance for pressure variation. A $\pm 10\%$ variation is used for turbulent flow distributors and $\pm 5\%$ to $\pm 7\%$ for laminar flow distributors.

Step 10. Start at the closed end and select the smallest pipe or the Working Curve number. Keeping the vertical line on the tracing paper coincident with the ordinate axis of the underlying curves slide the tracing paper up and down and fixes the position that includes maximum length of the pipe within the parallel lines of the tolerance limit. Draw the length of the curve.

Step 11. Select the next pipe size or the working curve number. Similar to Step 10 slide up and down the tracing paper to include the maximu length of this pipe size within the tolerance limit. Draw this length of the curve on the tracing paper.

Step 12. The process will continue till the selected length of the pipes reach to the inlet end.

Step 13. Determine the length of different diameter pipes from the points of intersections of the curves drawn on tracing paper for different pipe sizes.

Step 14. From the pressure distribution curves so prepared, determine q_{min} and q_{max} and calculate E_u for given M_r.

Example 4.3 A lateral of length 150m is to supply the water to the plants 3m apart at the rate of 5 l/h from the distributors. The characteristics of the distributors are q = $0.79H^{0.8}$, $CV = 4\%$ Lateral is to run a fairly uniform slope of about 1.0% down. Normal design pressure is 10m of water. Available ID of pipes are 10mm, 12mm and 15mm. Design the lateral by polyplot technique. Also find the and q_{min} and E_u.

Solution : $\text{Step1.SDR} = \frac{\text{Discharge per distributor}}{\text{Discharge between the distributors}} = \frac{5}{3} = 1.67 l/h/m$

Step 2-4. For ID 10mm → Working Curve No.17

„ ID 12mm → „ „ No. 22

„ ID 15mm → „ „ No.26

Step 5. Working Curve Nos. 17, 22 and 26 are marked on the main chart.

Step 6. Nominal design pressure, H=10m of water.

Step 7. Ordinate and abscissa lines are drawn on the tracing paper following the main chart.

Step 8. Using both ordinate and abscissa from the closed end a line is drawn showing the actual profile of the lateral (1.0%) over which the lateral is to run.

Step 9. Parallel to the profiles, lines are drawn showing ±5% tolerance i.e., $\pm 0.5m$.

Step 10-13. Keeping the ordinate fixed and sliding the tracing paper up and down the portion of the curves within the tolerance zone are drawn for different diameters. From the point of intersection of the curves the lengths are selected 48, 59 and 43m for 10, 12 and 15mm diameters respectively.

Step 14. From Fig. 4.10 (curves on tracing paper) it is found that total elevation and friction head, $\Delta h_{total} = \Delta h + \Delta h_{el} = 2.0m$. But, $\Delta h_{el} = 1.5m$. So, $\Delta h = 2.0 - 1.5 = 0.5m$.

Head, $H(i = 1) = 10m - 0.5m(5\%) = 9.5m$

$H(i = 0) = 9.5m + 0.5m = 10.0m$

$q = 0.79H^{0.8}$

$q_{min} = 0.79(9.5)^{0.8} = 4.78 l / h$

$$\bar{q} = 0.79(10)^{0.8} = 4.98 l/h$$

$$Mrf(e) = 1 - \frac{1.27}{\sqrt{e}}.CV = 1 - \frac{1.27}{\sqrt{1}} - \frac{4.0}{100} = 1 - 0.0508 = 0.949$$

$$Eu = 100.\frac{q\min}{\bar{q}}.M_r f(e)$$

$$= 100.\frac{4.78}{5.00}.0.949$$

$$= 91.09\%.$$

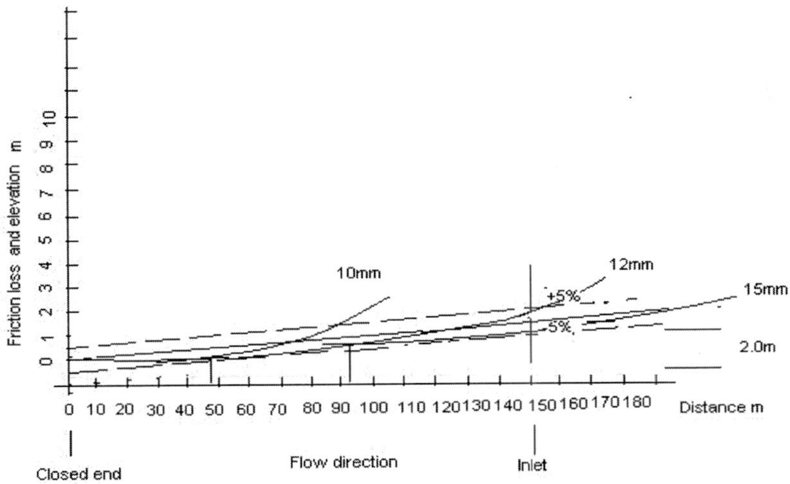

Figure 4.10: Graphical design of pipe network-case I

Case II. Fixed outlet discharge with constant pipe diameter throughout

The design procedure is same to Case-I excepting that there should be one acceptable pipe diameter throughout instead of a few. It is to be tried to find out the lowest number curve, which suits within the profile for the entire length of the lateral (Fig. 4.11). For the Example 4.3 it appears that curve number 23 will be best fit. The diameter of the pipe is now selected by using this curve number for given SDR from the top left-hand graph of Fig. 4.9. It is the 12mm diameter pipe for Example 4.3.

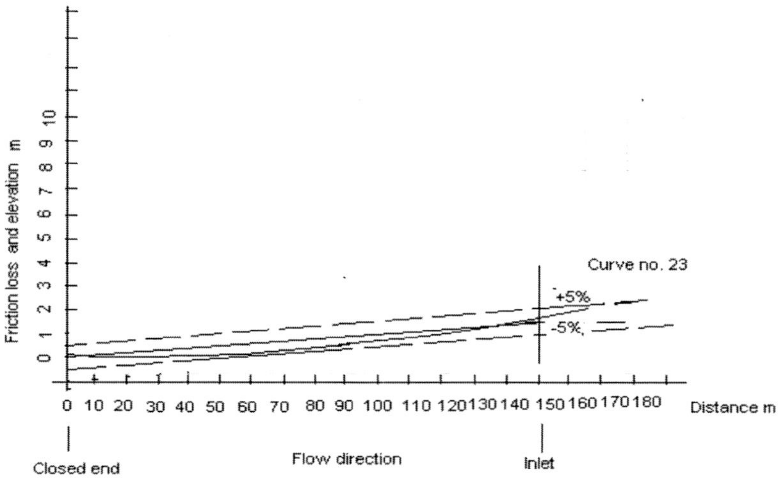

Figure 4.11: Graphical design of pipe network-case II

Case III. Outlets that can compensate pressure variation and constant pipe diameter throughout

The outlets are micro tube. By using different length of the micro tube the effects of pressure variation may be adjusted. For this purpose the **length/ pressure** curve of micro tube is to be known. Fig. 4.12 is having such a curve for the outlet discharge.

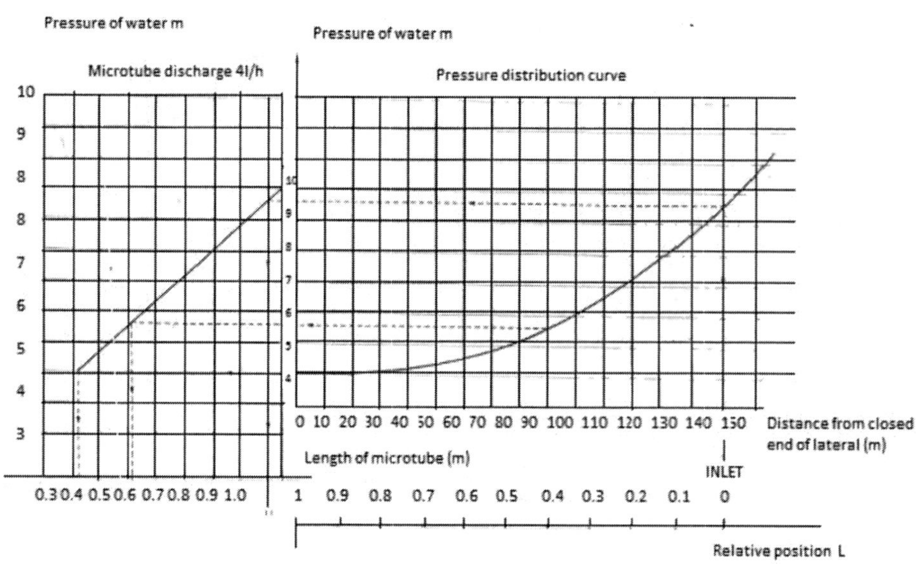

Figure 4.12: Graphical design of pipe networks

Sub-main Design

Distribution of pressure within the unit of sub main pipe network and emission uniformity.

The position of a distributor in a lateral of a sub-main may be characterized by a couple (M, L). The M and L represents the relative position of the sub main and lateral respectively. The relative position at the inlet is 0.0 and 1.0 at the far end of sub main or the laterals. Thus the pressure head at any position can be noted as H (M, L). Soloman and Keller (1974) proposed the formulae to calculate H (M, L) at all the points in the network knowing only the head at the inlet of the sub main H(0,0), the head at the downstream end of the downstream lateral H(1,1) and the head at upstream end of the downstream lateral H(1,0) (Fig. 4.13). The formulae are given for following cases:

Sub main design with constant ID pipe-flat field

(4.32)

The E (L) and E (M) values can be obtained from Table 4.5. L and M is to indicate the relative position or length ratio of lateral and sub-main respectively. H (0, 0); H (1, 0); H (1, 1) are known (Fig. 4.13). X/x is ratio of lateral discharge exponent X to the distributor discharge exponent x. The values of ratio X/x are given in Fig. 4.14 for various values of x and $\Delta H / \bar{H}$ of the last lateral on the sub main.

$$\frac{\Delta H}{\bar{H}} = \frac{H(1,0) - M(1,1)}{0.23H(1,0) + 0.77\,H(1,1)}$$

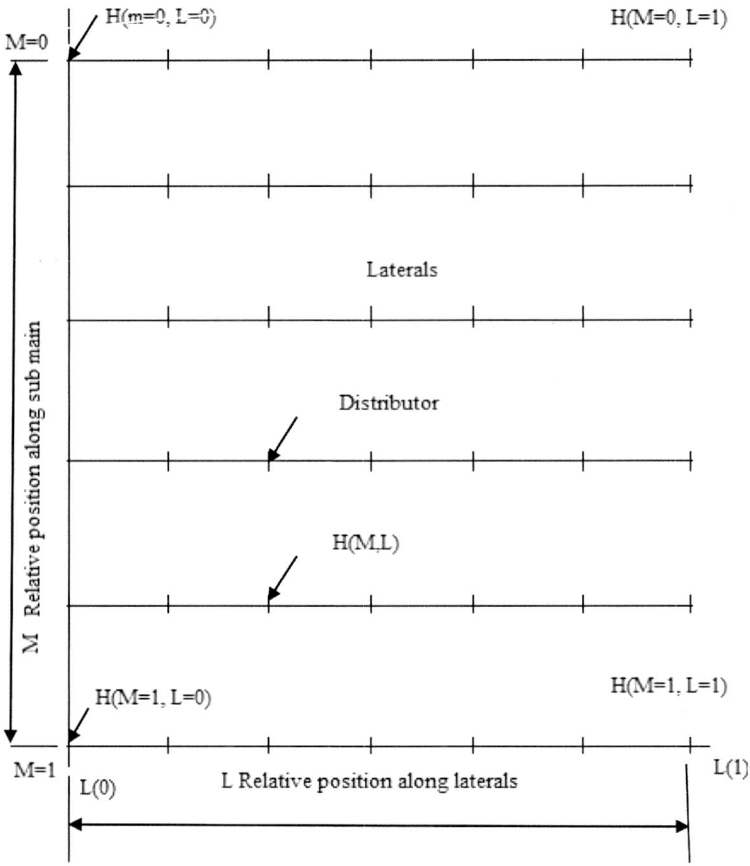

Figure 4.13: A drip system sub unit

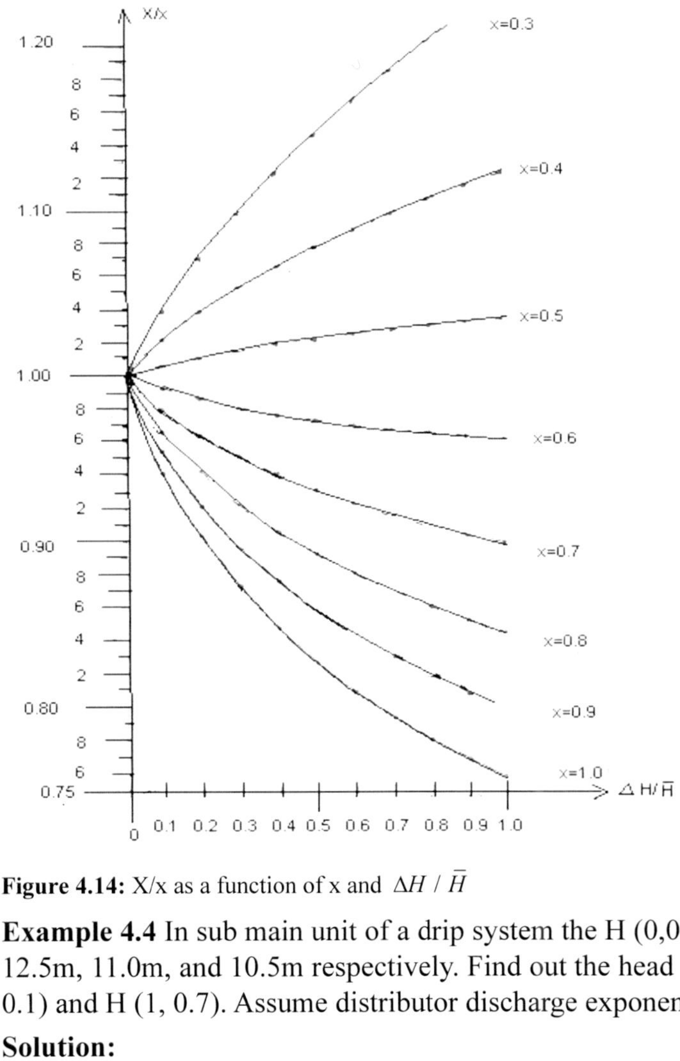

Figure 4.14: X/x as a function of x and $\Delta H / \bar{H}$

Example 4.4 In sub main unit of a drip system the H (0,0), H (1,0), H (1,1) is 12.5m, 11.0m, and 10.5m respectively. Find out the head H (0.2, 0.2), H (0.5, 0.1) and H (1, 0.7). Assume distributor discharge exponent as 0.80.

Solution:

$$\frac{\Delta H}{\bar{H}} = \frac{H(1,0) - H(1,1)}{0.23H(1.0) + 0.77\,H(1,1)}$$

$$= \frac{11.0 - 10.5}{0.23 \times 11 + 0.77 \times 10.5}$$

$$= \frac{0.5}{2.53 + 8.085} = \frac{0.5}{10.615} = 0.047$$

$$\text{From fig. 4.14, } \frac{X}{x} = 0.98 \text{ for } \frac{\Delta H}{\bar{H}} = 0.047$$

Eq. 4.32,

$$\mathrm{H(M,L)=[1.3E(L)-0.3][E(M)H(0,0)+\{1-E(M)\}H(1,0)]}$$
$$\mathrm{+[1-E(L)][H(1,1)+0.3H(1,0)]}$$
$$\left[\mathrm{E(M)\frac{H(0,0)}{H(1,0)}+(1-E(M)}\right]^{\frac{X}{x}}$$

$$H(1,0.7)=[1.3\times 0.093-0.3][0.00\times 12.5+(1-0.00)11.0]$$
$$+[1-0.093][10.5+0.3\times 11.0]\left[0.0\times\frac{12.5}{11.0}+(1-0.0)\right]^{0.98}$$
$$=[-0.179][11.0]+[0.907][13.8][1]^{0.98}$$
$$=-1.97+12.517$$
$$=10.547m$$

Completely tapered sub main

If the head loss per unit length is constant;

$$H(M,L)=[1.3E(L)-0.3][(1-M)H(0.0)+M.H(1,0)] \quad (4.33)$$
$$+[1-E(L)][H(1,1)+0.3H(1,0)]\left[(1-M)\frac{H(0,0)}{H(1,0)}+M\right]^{\frac{X}{x}}$$

Example 4.5 Find the head H (0.3,0.3) and H (0.6,0.6) in a completely tapered sub-main in flat field assuming H (0,0), H (1,0) and H (1,1) as 14.5m, 12.5m and 12.0m, respectively. The distributor discharge exponent is 0.8.

Solution

$$\frac{\Delta H}{\overline{H}}=\frac{H(1,0)-H(1,1)}{0.23H(1,0)+0.77H(1,1)}$$
$$=\frac{12.5-12.0}{0.23\times 12.5+0.77\times 12}$$
$$=\frac{0.5}{12.115}=0.04127$$

For $\frac{\Delta H}{\overline{H}}=0.04127$, $X/x=0.982$ from Fig. 4.14.

Using Eq. 4.33,

$$H(0.3,0.3)=[1.3\times0.64-0.3][(1-0.3)14.5+0.3\times12.5]$$

$$+[1-0.64][12.0+0.3\times12.5]\left[(1-0.3)\frac{14.5}{12.5}+0.3\right]^{0.982}$$

$$=[0.532][13.9]+[0.36][15.75][1.112]^{0.982}$$
$$=7.395+5.67[1.1099]$$
$$=7.395+6.293$$
$$=13.69m$$

$$H(0.6,0.6)=[1.3\times0.175-0.3][(1-0.6)14.5+0.6\times12.5]$$

$$+[1-0.175][12.0+0.3\times12.5]\left[(1-0.6)\frac{14.5}{12.5}+0.6\right]^{0.982}$$

$$=-0.0725+[14.4743][1.0628]$$
$$=-2.5895+13.81$$
$$=12.846m$$

Submainwith pressure regulator placed at the inlet of each lateral

$$H(M,L)=H(1,1)+E(L)[R-H(1,1)] \tag{4.34}$$

Where, R=constant output head of the regulator.

Example 4.6 Find the head H (0.5, 0.5) in sub-main drip network where the regulators are used at each of the lateral at the output head of 12.0m. The H (1, 1) is 11.4m.

Solution

$$H(M,L)=H(1,1)+E(L)[R-H(1,1)]$$
$$or,\ H(0.5,0.5)=11.4+0.297[12.0-11.4]$$
$$=11.4+0.178$$
$$=11.58m.$$

Design Charts

The hydraulic calculation discussed in previous sections is tedious. The design charts have been developed by computer simulation to select the main, sub main and the laterals.

Lateral line design chart

a. For uniform slope

The 12 mm and 16 mm lateral lines sizes are commonly used in field practices. Fig. 4.15 and 4.16 represents the design charts for these two lateral sizes. Fig. 4.17, 4.18 and 4.19 represent the general charts that can be used for any pipe size. The design procedures are as follows:

Step 1. Establish the lateral length (L), operating pressure head (H), ratio L/H and determine the total discharge (Q) in liters per second (lps).

Step 2. Move vertically from L/H (in quadrant III) to the given total discharge (lps) in quadrant II, then establish a horizontal line in quadrant I.

Step 3. Move horizontally for L/H (quadrant III) to the percent slope line in quadrant IV, then establish a vertical line in to quadrant I.

Step 4. The intersection point of these two lines in quadrant I determines the acceptability of the design.

Desirable: Pressure variation less than 20% or emitter flow variation less than 10%.

Acceptable: Pressure variation 20-40% or emitter flow variation about 10-20%.

Not recommended: Pressure variation greater than 40% or emitter flow variation larger than 20%.

Example 4.7 Select the lateral size of a drip system for the following operating condition:

Length of the lateral=75m

Operating pressure head=10m

Emitter discharge=4l/h

Spacing of emitters=0.75m

Slope of the lateral line=2 percent down slope

Solution: L=75m, H=10m

Step 1. Calculating, L/H=75m/10m=7.5

Total discharge in lateral, $Q = \frac{75m}{0.75m} x4l/h = 400l/h$

$= 0.111l/s$

Step 2. Using Fig. 4.16 and moving vertically from L/H=7.5 to the total discharge line (Q=0.111 l/s), the horizontal line is established in to quadrant I.

Step 3. Moving horizontally from L/H in quadrant III to the 2% down slope in quadrant IV, the vertical line is established in quadrant I.

Step 4. The point of intersection of the horizontal line from quadrant II and vertical line from quadrant IV in quadrant I is in the desirable zone in which the pressure variation is found within 20% (and emitter flow variation is within 10%).

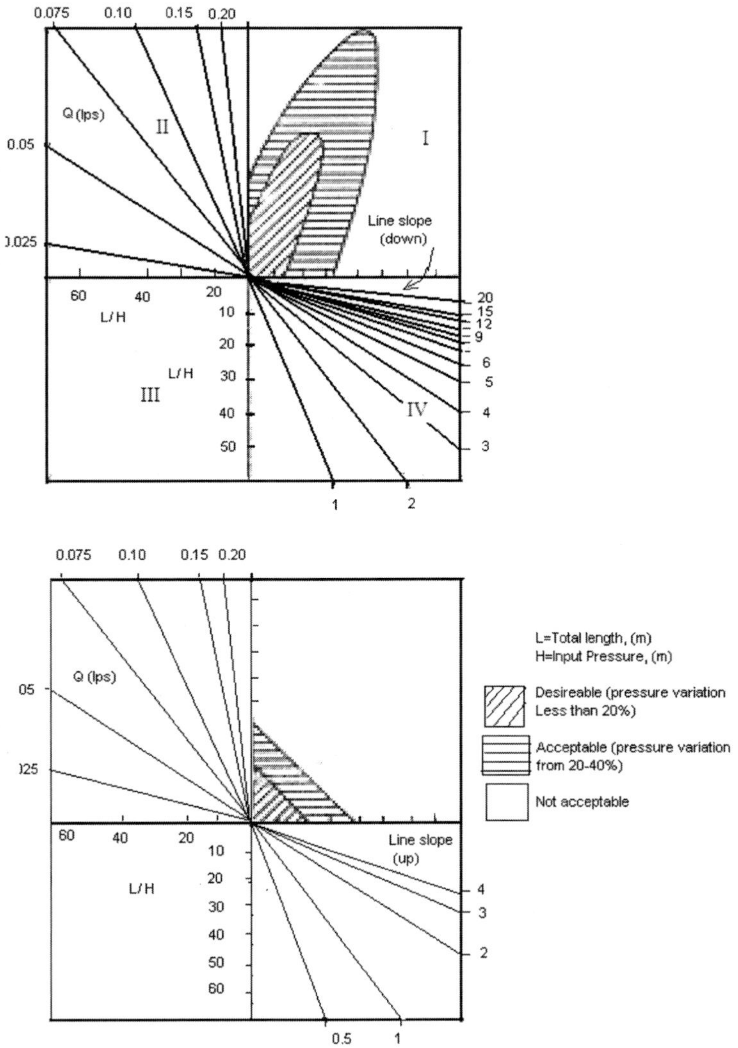

Courtesy: Michael (1984)

Figure 4.15: Design chart for drip lateral (12mm)

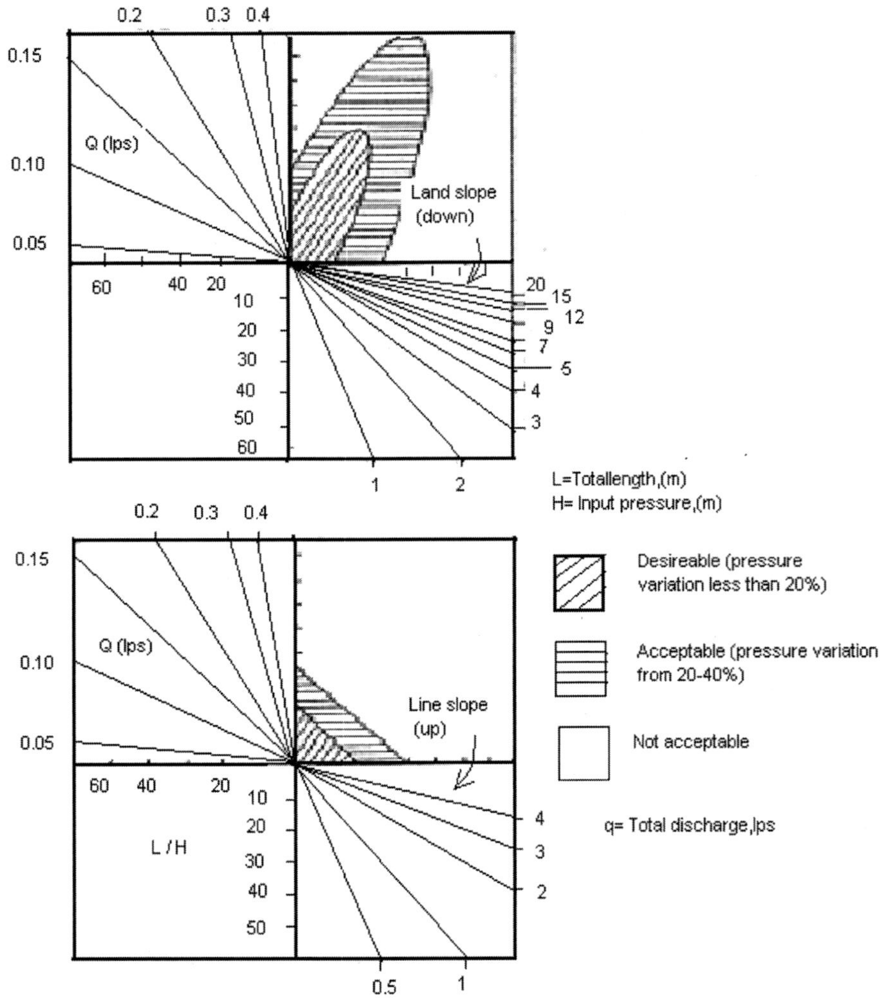

Courtesy: Michael (1984)

Figure 4.16: Design chart for drip lateral (16mm)

b. General design charts

Fig. 4.17 and 4.18 are plotted dimensionless for uniform down slope and up slope respectively. These can be used for any diameter pipes. Fig. 4.19 is a nomograph for determining $\Delta H/L$ from total discharge, Q, and pipe size, D. Fig. 4.17 to 4.19 can be used to check the acceptability of a design when the lateral size is given or to select a proper size of a lateral line in a given design condition. The design procedure is as follows:

i. To check acceptability of design when the lateral size is given.

Step 1. Establish a trial L/H and total discharge Q (lps).

Step 2. Determine $\Delta H / L$ from Fig. 4.19 by using the total discharge (Q) and lateral size (D).

Step 3. Move vertically from L/H in quadrant III to the determined $\Delta H / L$ in quadrant II of the appropriate figures (Fig. 4.17 or 4.18), then establish a horizontal line in quadrant I.

Step 4. Move horizontally from L/H to the % slope line in quadrant IV, then establish a vertical line in quadrant I.

Step 5. The point of intersection of the two lines in quadrant I determines the acceptability of the design.

Desirable: Emitter flow variation less than 10%

Acceptable: Emitter flow variation less than 20%

Not recommended: Emitter flow variation greater than 20%

ii. To select proper lateral size

Step 1. Establish L/H and lateral discharge, Q (lps)

Step 2. Move horizontally from L/H (in quadrant III) to the percent slope in quadrant IV of the appropriate figure (Fig.4.17 or 4.18), then establish a vertical line in to quadrant I.

Step 3. Select a point along this line in quadrant I at the upper boundary of the desirable region A or acceptable region B depending on the design criterion. From this point establish a horizontal line in quadrant II.

Step 4. Establish a vertical line in quadrant II from the L/H value. This line intersects at a point with the horizontal line drawn from quadrant I in step 3. Determine the $\Delta H / L$ value at this point.

Step 5. From the nomograph (Fig.4.19) using the total discharge and the $\Delta H / L$ value, determine the lateral discharge.

Example 4.8 Determine the size of the lateral for the lateral length of 100m in a crop field of 1% down slope. The emitters are spaced 0.5m apart and discharge at a rate of 4 l/h at the average pressure head of 10m.

Solution: Lateral length, L=100m

Operating pressure, H=10m

Field down slope=1%

Step 1. Calculating, L/H=100m/10m=10

$$\text{Total discharge, } Q = \frac{100m}{0.5m} x4l/h = 800l/h$$
$$=0.222 \text{ l/s}$$

Step 2-4. Horizontal movement is done with L/H=10 from quadrant III to 1% down slope in quadrant IV (Fig.4.17). From the point of intersection to down slope line in quadrant IV vertical line is drawn in quadrant I to select a point at the extreme end of desirable zone and the horizontal line is drawn to quadrant II from this point. Again the vertical line is drawn with L/H=10 in quadrant II which intersects at $\Delta H / L = 3.5\%$ with the horizontal line from quadrant I. Using the Fig.4.19 the lateral size found is approx. 16mm.

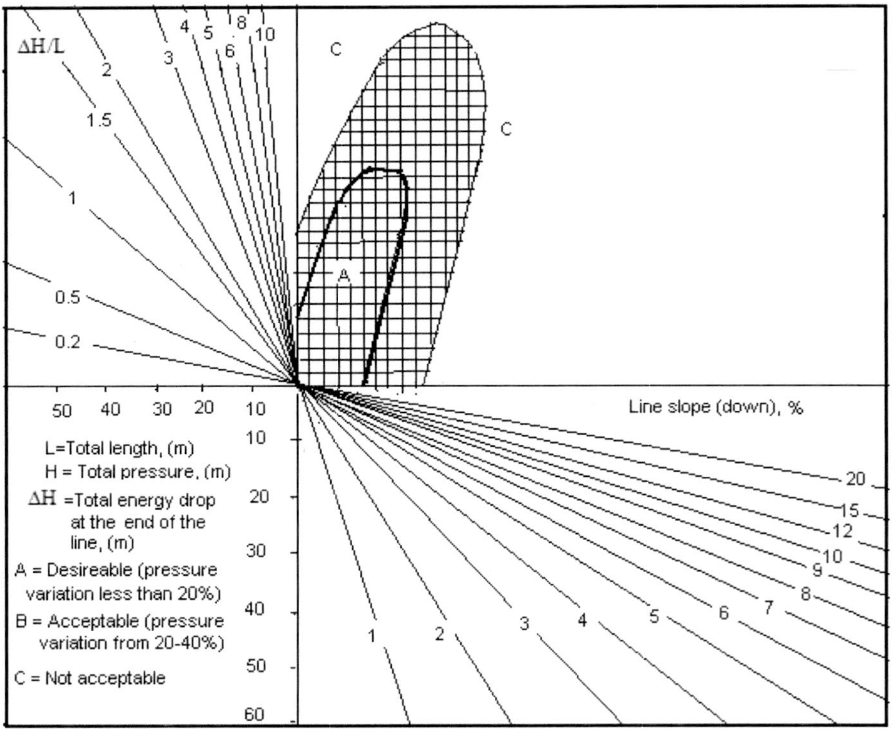

Figure 4.17: Dimensionless design chart (down slope)

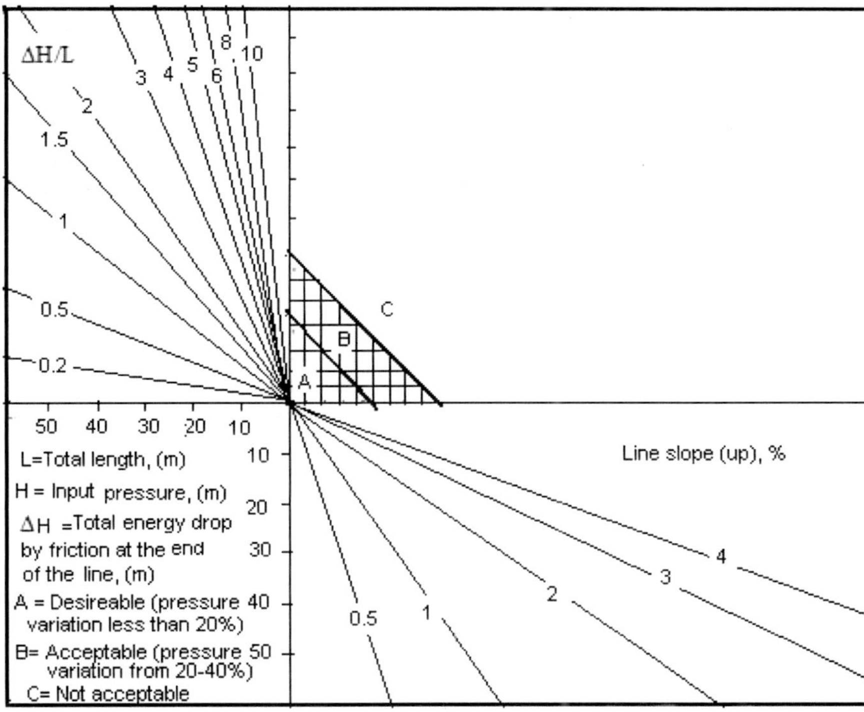

Figure 4.18: Dimensionless design chart (up slope)

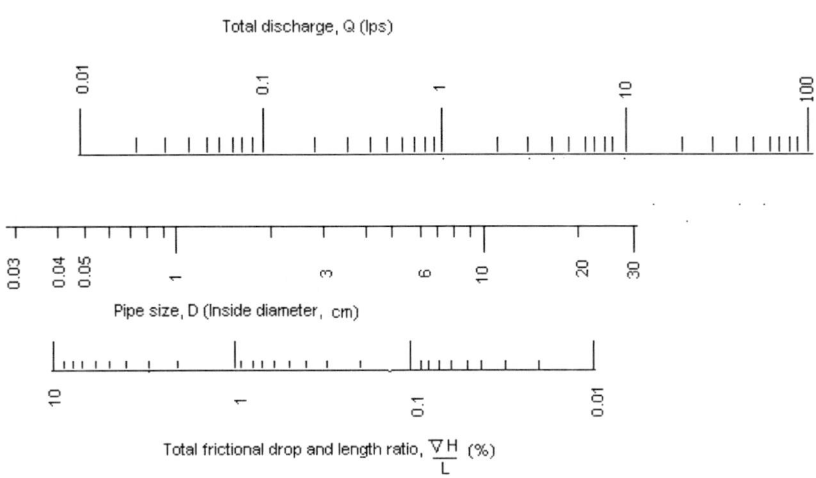

Figure 4.19: Nomograph for drip laterals and sub main design

Lateral design for non-uniform slope

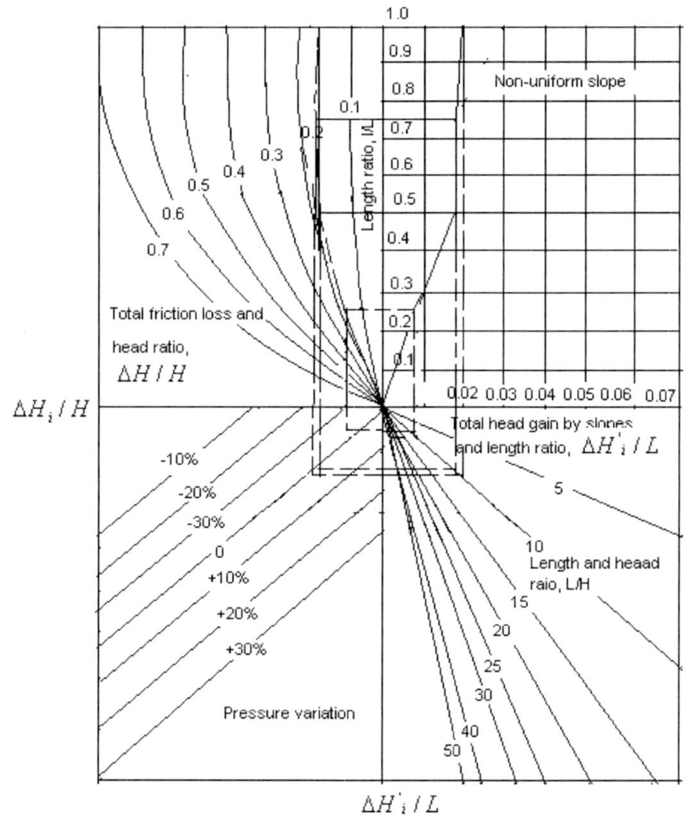

Figure 4.20: Design chart for non-uniform slope

This is a dimensionless chart developed by Wu and Gitlin (1975), which can be used for both metric and FPS units. The design procedure follows:

Step 1. Divide the non-uniform slope profile in to several sections each of which may be considered as uniform slope. Calculate the energy loss and gain by knowing the slope of each section. Determine the total energy gain due to slopes for any length along the line $\Delta H'_i$

Step 2. Plot the non-uniform slope pattern in a dimensionless form l/L i.e., the length ratio i (l/L) vs total energy gain due to slopes $\left(\Delta H'_i / L\right)$ in quadrant I (Fig.4.20).

Step 3. By using Eq. 4.9 determine the total energy drop (ΔH) by friction and calculate the total energy loss (ΔH) and operational pressure head (H) ratio $\left(\Delta H / H\right)$.

Step 4. Determine the lateral length (L) & operating pressure head (H) ratio L/H.

Step 5. Select a point from the non-uniform slope profile in quadrant I preferably point between two slopes or at any desired length ratio and establish horizontal line to the determined $\Delta H_i^{'}/H$ in quadrant II and establish vertical line to quadrant III.

Step 6. From the point selected in step 5 in quadrant I draw a vertical line downward to the determined L/H in quadrant IV and vertical line from quadrant II in quadrant III gives the pressure variation from the operating pressure.

Step 7. The location of intersection of horizontal line from quadrant IV and vertical line from quadrant II in quadrant III gives the pressure variation from the operating pressure.

Step 8. Repeat the entire process for some other points preferably at regular interval in the dimensionless non-uniform slope profiles and check the pressure variation of these points.

Example 4.9 A 100m long and 15mm diameter lateral line is laid to uneven land of 3% down, 4% down, 0% and 1% down for 0-25m, 25-50m, 50-75m and 75-100m length respectively. The total discharge through the lateral is 0.13l/s. Assume 10m as the operating pressure head for the drip system. Check the pressure variation in lateral.

Solution: Given

$$L = 100m$$

$$H = 10m$$

$$d = 15mm,\ q = 0.13l/s$$

$$\text{Length ratio, } l/H = 0.25, 0.5, 0.75, 1.0$$

Step 1&2. The total energy gain at different length ratio is calculated as follows:

l/L	$\Delta H_{i}^{'}, m$	$\Delta H_i^{'} / L$
0.25	0.75	0.0075
0.50	1.75	0.0175
0.75	1.75	0.0175
1.00	2.00	0.02

The non-uniform slopes in dimensionless forms and the respective energy given $\left(\Delta H_i^{'}/L\right)$ are shown in quadrant I (Fig.4.20).

Step 3&4. Total energy drop due to friction following Eq. 4.9,

$$\Delta H = 5.35\left(\frac{q^{1.852}}{d^{4.871}}\right)L$$

$$= 5.35\left\{\frac{(0.13)^{1.852}}{(1.5)^{4.871}}\right\}\times 100 = 5.35\times\left(\frac{0.023}{7.207}\right)\times 100 = 5.35\times 0.0032\times 100$$

$$= 1.7m$$

$$\therefore \Delta H/_{H} = \frac{1.7}{10} = 0.17$$

$$L/_{H} = \frac{100}{10} = 10$$

Step 5-8. Selecting the point at the length ratio 0.25 where $\Delta H_i'/_{L} = 0.0075$, horizontal line is drawn in quadrant II up to point of intersection of line represents $\Delta H/_{H} = 0.17$. From the selected point in quadrant I line is drawn vertically downward to quadrant IV to intersects with the line which represents $L/_{H} = 10$. From this point horizontal line is drawn to quadrant III. Vertically downward line is also drawn from the point of intersection in quadrant II to III, which intersects with horizontal line drawn from quadrant IV at the point to represents the pressure variation 7.5% negative. Similarly the pressure variation at 0.5, 0.75 and 1.0 length ratio are found 2.5%(+), 1%(+) and 3%(+), respectively.

Sub main design chart

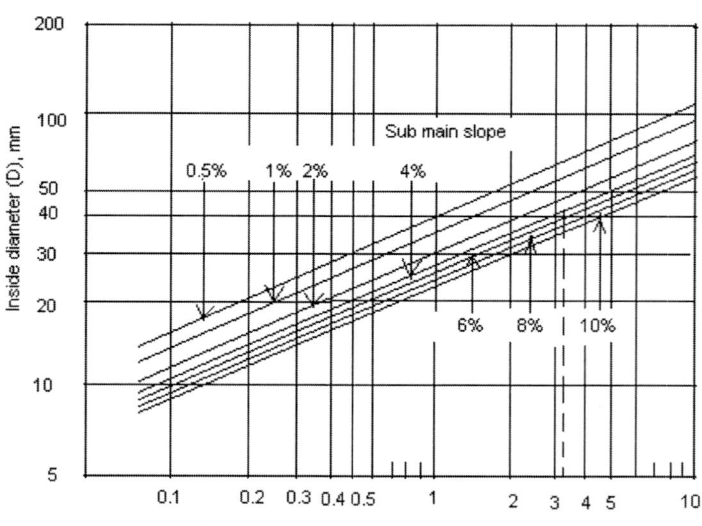

Figure 4.21: Sub main design chart (slope equal to or larger than 0.5%)

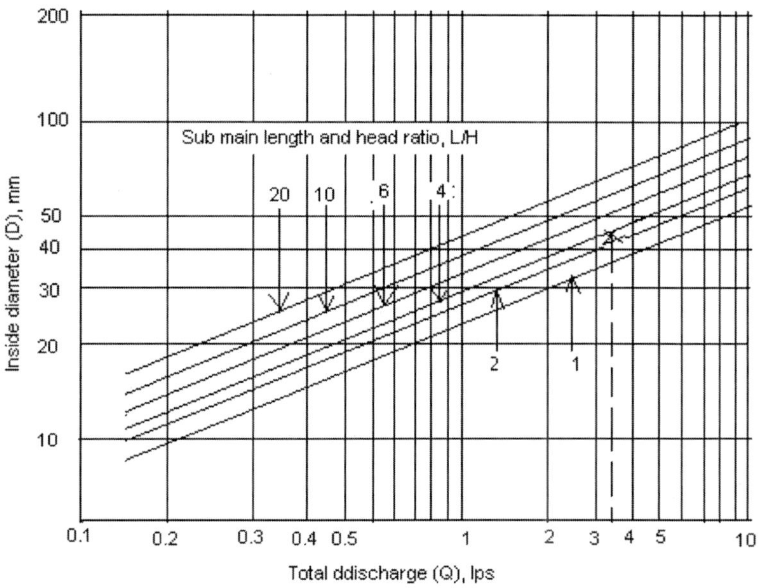

Figure 4.22: Sub main design chart (slope less than 0.5% and allowable pressure variation 10%)

The flow characteristics of a sub main and the laterals are same excepting the sub main size is larger to accommodate the entire flow of the laterals in a sub main. The design procedure using the charts as described in earlier sections can be used for designing sub main.

The sub main length is usually short (approx. 25 to 75m). It can be assumed that the total frictional drop is equal to total energy gain due to slope, $\Delta H = \Delta H'$. When the ΔH is made equal to $\Delta H'$, the Eq. 4.6 can be rewritten as

$$\Delta H' = 3.98 \times 10^5 \frac{q^{1.852}}{d^{4.871}} L$$

$$or, \quad S_0 = 3.98 \times 10^5 \frac{q^{1.852}}{d^{4.871}} \qquad (4.35)$$

where, S_0 is simply the slope of sub main. In the assumption of $\Delta H = \Delta H'$. or when using Eq. 4.35, the maximum pressure loss will be at the middle section whose magnitude is equal to $0.36\,\Delta H'$. Following the Eq. 4.35 chart is made as shown in Fig. 4.21 which is applicable for the slope equal to or larger than 0.5%. A slope less than 0.5% is considered level or zero. Therefore, the Eq. 4.35 cannot be used for such case. The pressure variation will be only due to frictional loss and it should be maximum at the end of the sub main. Therefore, the Eq. 4.6 can be rewritten as follows selecting the loss $\frac{\Delta H}{H} = 10\%$ maximum acceptable pressure variation.

$$\Delta H/_H = 3.98x10^5 \frac{q^{1.852}}{d^{4.871}} \frac{L}{H} \tag{4.36}$$

$$or\ 0.1 = 3.98x10^5 \frac{q^{1.852}}{d^{4.871}} \frac{L}{H} \tag{4.37}$$

The design chart as shown in Fig. 4.22 is made following the Eq. 4.37. The design procedure for using the simplified charts (Fig. 4.21 & 4.22) is follows:

Step 1. Determine the total discharge, Q, for the sub main.

Step 2. Determine the length and pressure ratio $L/_H$.

Step 3. Determine the sub main slope. If the slope is more than 0.5%, use Fig. 4.21. Use Fig. 4.22 for slope less than 0.5%.

Example 4.10 Using the lateral line as given in Example 4.8 with lateral discharge of 0.22 l/s, design a sub main when the sub main length is 30m and the lateral line spacing is 2m. Design the sub main size when sub main slope is zero and 4% uniform down slope.

Solution: Given:

Lateral line length=100m

Discharge per lateral size=0.22 l/s

Sub main length, L=30m

Operating pressure head=10m

Step 1. Number of lateral lines= $30/_2 = 15$

Total discharge of sub main inlet, Q=15x0.22=3.3 l/s

Step 2. $\frac{L}{H} = \frac{30}{10} = 3$

Step 3. When sub main slope is zero, using the Fig. 4.22, the sub main size is designed as 45mm. When the sub main slope is 4%, using Fig. 4.21, the designed sub main size is 42mm.

Main Line Design

Main line design is like a pipe design when it delivers water to one or two sub fields. In such case the size of main may be determined by using Eq. 4.5 $\left(\Delta H = 15.27 \frac{q^{1.852}}{d^{4.871}}.L\right)$. When the main delivers to many sub mains or sub fields the flow through it decreases with respect to length. The pipe size of the main at any section of it will depend on the energy gradient above the main line. The total energy at any outlet, along the main should be equal or higher than the energy required to operate the system under the command of that outlet. Therefore, the design approach mainly to determine the allowable energy drops for all main line sections. The design procedure is as follows:

Step 1. Plot the main line profiles as shown in Fig. 4.23.

Step 2. Plot the required pressure head along the main line profile as shown in Fig. 4.23.

Step 3. Draw the straight energy gradient line with reference to the required pressure profile such that the energy gradient line is along the required pressure profile (Fig. 4.23).

Step 4. Determine the energy slope which in the slope of the straight energy gradient line, $\Delta H/L$.

Step 5. Design the size of the main by using the nomograph (Fig. 4.19) based on the energy slope and the total discharge for each main section.

Example 4.11 A drip irrigation system is required for a 10ha field. The field is rectangular and divided by sub mains connected from the main line. The main line is laid on the center of the field with 10 sub plots on each side of the main line. Each sub plot is about 125m long and 40m wide and each main line section is 40m long. The design capacity is 1.5 l/s to each sub plot. There are a total of 9 outlets on each side of the main line and at the end of each section to supply 1.5 l/s to each sub plot. If the main line slopes, the required operating pressure (7.5m) at the lateral lines and available input pressure (18m) at the beginning of the main line are given in Fig. 4.23, design the main line.

Solution: Given, Main line length = 40x9 = 360m

Operating pressure for drip laterals=7.5m

Available input pressure at the inlet=18.0m (point A)

Total discharges to main = 30 l/s

Required total energy at the last field= 10.5m (point B)

Step 1. Main line profile is plotted as shown in Fig. 4.23.

Step 2. Required pressure head of 7.5m is plotted along the main line as shown in Fig. 4.23.

Step 3. Straight energy gradient is drawn and the energy slope is determined as 2.08%.

Step 4. By using the nomograph (Fig. 4.19) and energy slope 2.08% the main line size at different sections are determined and presented in Table 4.8.

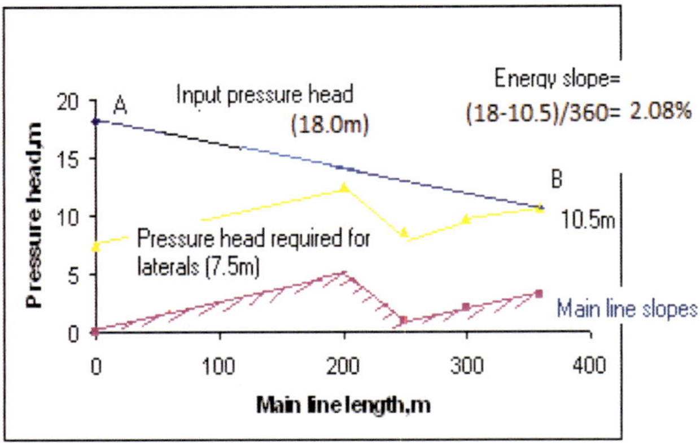

Figure 4.23: Main line lay out and energy gradient line

Table 4.8: Main line size determined from nomograph (Fig. 4.19) for Example 4.11.

Main line section	Discharge,l/s	Main line size inside dia, cm
0*	30	15
1	27	12.5
2	24	12.5
3	21	12.5
4	18	12.5
5	15	12.5
6	12	12.5
7	9	10.0
8	6	7.5
9	3	7.5

*There is an inlet at the entrance of section 1 for irrigating the sub plots on both side of the section

Farm Drip System Design Examples

Example 4.12 Design the drip irrigation system for banana crop grown in 1ha field at 2mx2m spacing in New Alluvial Agro-climatic zone of West Bengal

Solution

Layout of the field: Let the length & breadth of the field = 100mx100m

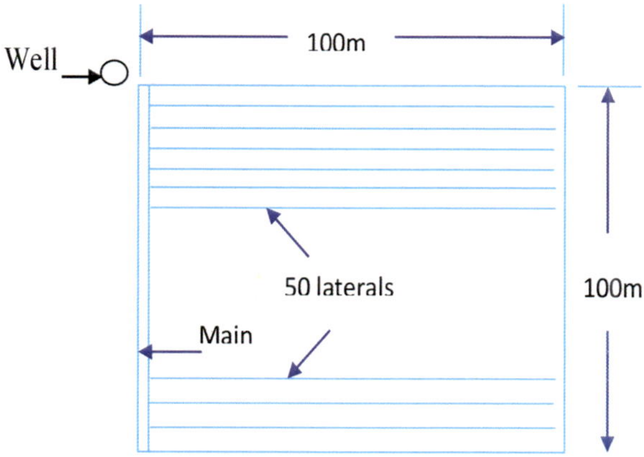

Figure 4.24: Layout of field for 1ha banana crop

Water requirement of crop:

Water requirement, $ET_{crop} = E_{pan} x K_p x K_c x A_w x S_p$

Where, ETcrop= water requirement/plant/day

E_{pan} = pan evaporation, mm/day

K_p = pan factor

K_c = crop factor

A_w = wetted area factor (0.3 for wide spaced crops and 0.9for closely spaced crops)

S_p = spacing of crops, m²/plant

For New – Alluvial Agro-climatic zone, the average of the maximum evaporation of 2 weeks in summer = 6.5 mm (say)

$\therefore E_{pan} = 6.5$ mm/day

$K_p = 0.8$

$K_c = 1.1$ (Anonymous)[1]

$A_w = 0.5$ (assumed)

$S_p = 2\text{m}x2\text{m} = 4\text{m}^2$

$$ET_{crop} = 6.5mm / day \times 0.8 \times 1.1 \times 0.5 \times 4m^2$$
$$= 11.44 \times 10^{-3} m^3 / day$$
$$= 11.44 l / plant / day$$

Total number of plants in the field $= \dfrac{Area}{Spacing} = \dfrac{10,000m^2}{4m^2 / plant} = 2500$

Daily water requirement of the area=11.44l/*plant*/*dayx* 2500=28600l=28.6m^3

Selection of drippers:

Say, the drippers to be used of capacity 2l/h. The actual discharge rate of drippers or the rate of application to a plant may be decided based on wetting pattern to the soil condition.

Time of operating or pumping the system $= \frac{11.44}{2} = 5.72h$

Alternately, a reservoir of 28.6m3 capacity may provide uninterrupted irrigation. In the area where power cut is the regular feature the provision of water reservoir may provide the scope of undisturbed irrigation.

Rate of pumping $= \frac{28600}{5.72} = 1.39lps$

Main line and laterals

Main

The size of the plot is assumed 100mx100m. Therefore, the main line would be length of 100m. The laterals are also of each 100m length originated from the main. The number of laterals would be 100m/2m = 50m. The total number of drippers per lateral = 100m/2m = 50.

Main line discharge = 1.39lps

Friction head loss in main pipes (m):

Total length = 100m

Equivalent length of 17 straight connection = 8.5m (assuming 6m pipe piece & 0.5m equivalent length for friction to each joint)

Bends & other fittings = 5.0m

Total main length = 113.5m

By using the Hazen-William formula where flow decreasing to zero at the end of the pipe,

$$h_l = 3.98x10^5 x \frac{q^{1.852}}{d^{4.871}} xL,$$

Where, Hazen-William constant C=150 for PVC.

q = flow in l/s

d = internal diameter of pipe in mm

Assuming the diameter of pipe = 50mm

$$h_l = 3.98x10^5 x \frac{1.39^{1.852}}{50^{4.871}} x113.5$$

$$or, h_l = 3.98x10^5 x \frac{1.84}{1.886x10^8} x113.5$$

$$or, h_l = \frac{831.27}{1.886x10^8} x10^5$$

$$\therefore h_l = 0.44m$$

Assuming 40mm diameter of pipe,

$$h_l = 3.98x10^5 x \frac{1.39^{1.852}}{40^{4.871}} x113.5$$

$$or, h_l = 3.98x10^5 x \frac{1.84}{6.36x10^7} x113.5$$

$$= 1.31m$$

The loss of head in friction also can be had from the chart for different diameter and flow.

Design criteria

1. The friction head loss in main should not exceed 1m/100m length
2. The head loss in a lateral should be such that it should not exceed 10% of the head at first emitter or average operating pressure usually assumed 10m.

In consideration to the above, the diameter of pipe for main to be selected is 50 mm.

Laterals

No. of dripper per lateral = 100m/2m = 50

Flow through each lateral $= 50x2 = 100lph = 0.0277lps$

In 100m length of the lateral there are 50 drippers inserted in to the lateral. The friction loss usually taken 0.5m length of lateral equivalent to a dripper. Thus equivalent length for 50 drippers in a lateral = 25m and the total lateral length to be considered = 100+25 = 125m

Using Hazen-William equation,

$$h_l = 3.98x10^5 x \frac{q^{1.852}}{d^{4.871}} xL$$

$$= 3.98x10^5 x \frac{(0.0277)^{1.852}}{(12)^{4.871}} x125$$

$$= 3.98x10^5 x \frac{1.31x10^{-3}}{180589} x125$$

$$= 3.98x10^5 x \frac{1.31x10^{-3}}{(12)^{4.871}} x125$$

$$= 3.98x10^5 x7.26x10^{-9} x125$$

$$= 0.36m$$

For 10mm lateral,

$$h_l = 3.98x10^5 x \frac{(0.0277)^{1.852}}{(10)^{4.871}} x125$$

$$= 0.87m$$

Assuming standard average operating head in lateral, the permitted head loss is 10/100 = 1m. The 12mm lateral is thus providing the friction loss much less than the permissible limit. The head loss in 10mm pipe is close to the extreme limit of 1.0m. Due to deposition of foreign particles in the system or the development of algae and fungi there is every scope of exceeding the permissible limit of friction head loss in the lateral at any time. It is; therefore, better to select the 12mm diameter lateral to be in the safe side.

Horsepower of the pump set

The pump uses to operate under static and friction head.

Static head

Depth of water	= 10.0m (say)
Draw down	= 2.0m (say)
Delivery head	= 1.0m (say)
Friction losses in pipes, bends, etc.	= 2.0m (say)
Total static head	= **15m**

The friction loss in the drip unit

Friction loss in the main = 0.44m

Friction loss in laterals = 0.36m

Head losses in control head for fertilizer applicator, filter, etc. = 10m

Minimum head required over dripper = 10.0m

Total head loss in drip unit = **20.80m**

Total head = Static head + Friction loss in drip unit

= 15+20.80=35.80m

Horsepower of the pump, $HP = \frac{QxH}{75xE}$

Where, Q = discharge, l/s

H = head of water, m

E = pumping efficiency

$$\therefore HP = \frac{1.39x35.80}{75x0.6} = 1.08$$

Efficiency may be assumed 60% for electric pump set and 40% for diesel pump set. Exactly 1.08hp motor may not be available in the market. Therefore, the next available size of the motor to be purchased.

The detail of the materials required for the crop field (1ha).

Sl. No.	Items	Quantity	Rate (Rs./-)	Amount (Rs)
1	50mm PVC pipe	160m	55/m	8,800
2	LDPE lateral pipe	5000m	7.5	37,500
3	Drippers (1l/h)	5000 pcs	3.5/pc	17,500
4	Sand filter	1no.	20,000	20,000
5	Fertilizer applicator (venturi assembly)	1no.	1,500	1,500
6	Ball valve	1no.	750	750
7	Flush valve (50mm)	1no.	450	450
8	Pressure gauge	1no.	650	650
9	Tees, joints, bends, etc. accessories [@10% of Sl. No.1-8 (Mane *et al.*, (2006)]	-	-	8,715
Total				95,865

Example 4.13 Design a drip irrigation system for 120m x 100m orchard field with the following.

Crop = Guava

Spacing = 5m x 5m

Soil = Sandy loam of saturated hydraulic conductivity 1.75m/day

Maximum evaporation = 6.5mm/day

Land slope = 0.1% (West to East)

Source of water = Well at corner of the field as shown in Fig. ..

Wetted area of crop = 30%

Crop factor & crop coefficient = 0.8

Location = New Alluvial Agro Climatic Zone of West Bengal

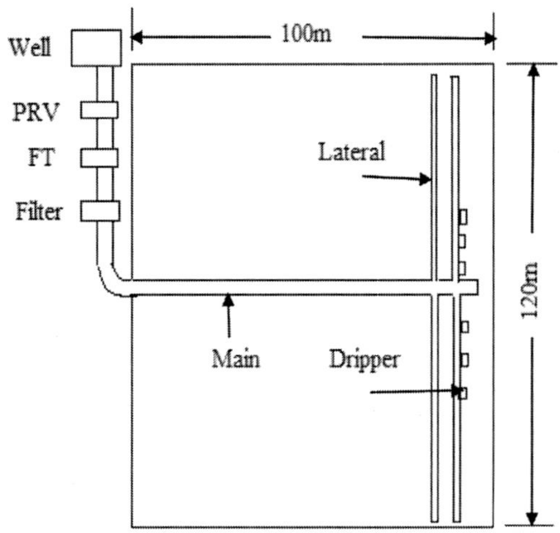

Figure 4.25: Layout of the field (100mx120m) for guava plant

Solution

1. Water requirement, $ET_{crop} = E_{pan} \times K_p \times K_c \times A_w \times S_p$

$= 6.5 \times 0.8 \times 0.8 \times 0.3 \times 5 \times 5$

$= 31.2$l/day/plant

With 90% irrigation efficiency the water requirement

$= 31.2/0.9 = 34.66 = 35$l/plant/day (say)

2. Selection of drippers

The plants are wide spaced. It is better to use a few drippers to wet more area for better root spreading and anchorage. Let 4 liter drippers 2 numbers per plant are used. Following Schwarzman & Zur , the depth of wetting ,

$$z = K_1 (V_w)^{0.63} \left(\frac{C_s}{q} \right)^{0.45}$$

$$= 29.2(35/2)^{0.63} \left(\frac{1.75/(24x3600)}{4} \right)^{0.45}$$

$$= 29.2x6.07x4.14x10^{-3}$$

$$= 0.73m$$

The depth of penetration may be accepted for guava plant for NAZ where ground water level is usually close to ground surface even in the summer. Here in the capillary rise of ground water is a continuous process.

3. Selection of main

Let the main enters into the field through the middle of N-S side and goes towards east following the natural slope. There should not be any sub main. Total plants in the area are to be irrigated at a time.

Length of the main = 60m + (100-5/2) =157.5m

No. of plants $= \frac{120x100}{5x5} = 480$

Discharge rate through the main = 480 x 4 x 2 = 3840l/h = 1.07l/s

Let the diameter of the main = 50mm

Using Hazen-William equation, head loss in the main,

$$h_l = 3.98x10^5 x \frac{(q)^{1.852}}{d^{4.871}} x157.5$$

$$= 3.98x10^5 x \frac{(1.07)^{1.852}}{(50)^{4.871}} x157.5$$

$$= 3.98x10^5 x1.13/(1.89x10^8)x157.5$$

$$= 0.38m$$

Friction loss in main is less than the recommended maximum loss of 1m. Therefore, the main diameter 50mm is accepted.

4. Selection of laterals

No. of plants in a lateral = 120/ (5x2) =12

No. of drippers in a lateral = 12 x 2 =24

The discharge in a lateral = 24x4 = 96l/h = 0.027l/s

Length of the lateral = 120/2-5/2 = 60-2.5 = 57.5m

Let us assume barb loss due to penetration of drippers in laterals is equivalent to 0.2m. length for each dripper [Biswas *et al.* (2005) & Reddy *et al.* (2001)]

Equivalent length of barb losses in a lateral = 24x0.2 = 4.8m

Let diameter of the lateral pipe = 10mm

Head loss in a lateral, $h_l = 3.98x10^5 x \frac{(q)^{1.852}}{d^{4.871}} x(L + L_e)$

$$= 3.98x10^5 x \frac{(0.027)^{1.852}}{(10)^{4.871}} x(57.5 + 4.8)$$

$$= 3.98x10^5 x1.67x10^8 x62.3$$

$$= 0.41m$$

The head loss 0.41m is less of recommended 10% head loss of average operating pressure (10m) i.e. 1m. The proposed 10mm diameter lateral pipe is accepted.

5. Selection of pump

Total head = static head (suction head+ delivery head) + average operating pressure of drip system + friction head loss in main line + friction head loss in lateral+ head losses in equipments in control head + head loss or gain due to elevation

Let, suction head = 12m, delivery head = 3m

Static head = 12 + 3 = 15m

Average operating pressure head of drip system = 10m

Friction head loss in main = 0.38m

Friction head loss in lateral = 0.41m

Let the fertilizer applicator, filter, pressure gauge, pressure relief valve, etc. are in use in drip system control head.

Let, head loss due to fertilizer applicator = 5m, filter = 2.5m,

pressure gauze, pressure relief valve and other fittings in control head = 2.5m

Head gain due to elevation difference = $\frac{0.1x97.5}{100} = 0.0975m \cong 0.1m$

Total operating head = 15 +10 + 0.38 + 0.41 +5 +2.5 +2.5 -0.1 = 35.7m

Horsepower of the pump, $HP = \frac{QxH}{75xE}$

$$= \frac{1.07x35.7}{75x0.6} = 0.85$$

Irrigation time at peak period $= \frac{35}{4x2} = 4.38 hours$

Example 4.14 The area of field and the crops and spacing are as shown below. Design the drip irrigation system for (i) entire area for banana and (ii) following different crops as shown.

Spacing of banana = 2mx2m

Spacing of papaya = 2mx2m

Spacing of orchard = 5m x5m

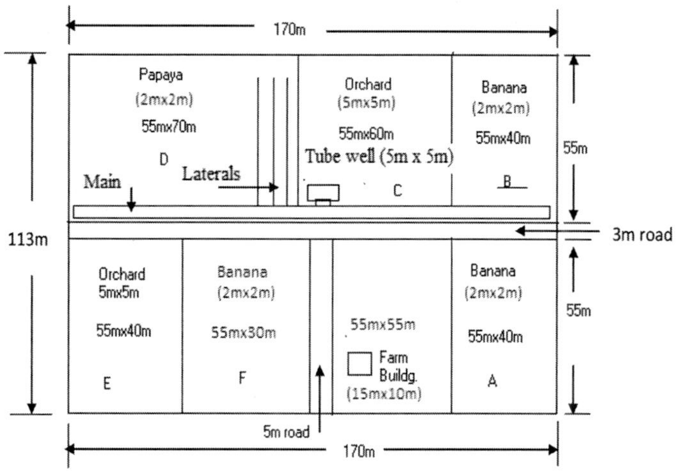

Figure 4.26: Lay out of the field (170m x 170m) for banana crop

Assuming the entire area is under banana cultivation

Length side = 170m

Width side = 55x2 +3m (road) = 113m

The main passes through the 3m road almost middle of the field.

Nos. of laterals on the main in each side = 170/2m = 85

Nos. of laterals on other side= (170-5)/2 =165/2=82.5≈83

Area not under cultivation:

Farm building = 15mx10m = 150m2

Main road = 5mx55m =275m2

Field road = 170mx3m = 510m2

Tube well = 5mx5m = 25m2

Misc. = 90m2

Total = 1050m2

Net cultivated area = 170mx113m-1050m2 = 18160m2

No. of banana plants = 18160/4 = 4540

Water requirement of the plants:

Pan factor(Kp) is taken 0.8. The pan evaporations (Epan), crop coefficients (Kc), wetted area (Aw) in different months as assumed are shown in Table 4.9.

Water requirement for the month of June,

$$V = E_p x K_p x K_c x A_w x S_p$$

=5.0x0.8x0.3x0.3x4=1.44l/plant/day

Assuming 30% wetted area in the months of initial stage of crop and 50% in development stage and onwards, crop period of banana 13 months and planting in the month of May, month wise water requirements of crop are tabulated in Table 4.9.

Table 4.9: Calculation of water requirement of crop in different months

Months	$E_{p,mm}$	K_p	K_c	A_w	Water requirement, l/plant/day	Irrigation interval, days	Water requirement, l/ plant/irrigation,
May	5.0	0.8	0.3	0.3	1.44	2	2.88
June	5.0	0.8	0.3	0.3	1.44	2	2.88
July	4.75	1.0	0.5	0.3	2.85	2	5.7
Aug	4.50	”	0.5	0.3	2.70	3	8.1
Sept	4.00	”	0.6	0.5	4.80	3	14.4
Oct	3.50	”	0.8	0.5	5.60	3	16.8
Nov	2.50	”	1.0	0.5	5.00	5	25.0
Dec	2.00	”	1.1	0.5	4.40	5	22.0
Jan	1.75	”	1.1	0.5	3.85	5	19.25
Feb	2.00	”	1.1	0.5	4.40	5	22.0
Mar	4.00	”	1.1	0.5	8.80	2	17.6
April	5.00	”	1.0	0.5	10.00	2	20.0
May	5.00	’	1.0	0.5	10.00	2	20.0

Irrigation interval

Irrigation interval may be selected on fixed deficit or fixed interval basis. Let us follow fixed interval basis such that the suction of soil before the irrigation should not exceed the permissible limit.

Following the suggestion of FAO (1980), the irrigation intervals in different seasons of the year may be as below.

Climate	Months	Irrigation interval, days	Irrigation interval proposed, days
Hot	March-July	2	2
Moderate	Aug-Oct	3	3
Cool	Nov-Feb	7	5*

- It has been stated that the drip irrigation in banana are usually made on daily basis or even in pulses several times/day irrespective of pan evaporation (http..). The region of NAZ, West Bengal is not extreme in behavior, rain occurs during July-September and even in winter, and also there is scope of moisture contribution from the soil profile since soil depth is quite high and the ground water is within few meter. Therefore, irrigation interval as suggested by FAO(1980) may be suitably modified for 5 days interval in November-February instead of 7 days.

Irrigation intervals and the water requirement/irrigation/plant are shown in Table 4.9.

Distributor to be used

It has been stated that nearly 88% of the root area located within 30cm of the soil surface and 97% within 40cm. It is recommended; therefore, that even under condition allowing unimpeded vertical root distribution, banana irrigation should be scheduled to wet only 30cm of soil depth (Anonymous)[1].

In the initial stage of crop growing when roots are in the process of growing the application of water to the maximum depth or width is not that important. Therefore, the penetration of water at the developing and mid stage may be examined. The representative volume of water applied per irrigation/day/plant is taken 14.4l following the Example 4.12.

Let the drippers to be used of capacity 2 l/h in single or in pair at either side of the plant.

Distribution of water at plant bottom:

For single dripper at the plant and 14.4l water volume per application,

Using Schwarzman & Zur (1985) [Eq. 2.6 & 2.7],

Depth of wetting front, $z = K_1 (V_w)^{0.63} \left(\frac{C_s}{q} \right)^{0.45}$

$$= 29.2(14.4)^{0.63}\left(\frac{0.25}{\frac{24x3600}{2}}\right)^{0.45}$$

$$= 29.2x5.367x2.356x10^{-3}$$
$$= 0.37m$$

The saturated hydraulic conductivity Cs is taken 0.25m/day for silty clay loam (Table 4.10).

Table 4.10: Saturated hydraulic conductivity (K) and drainable porosity (μ) values according to the soil texture

Texture (USDA)1	Structure	μ	K (m/day)
C, heavy CL	Massive, very fine or fine columnar	0.01-0.02	0.01-0.05
	With permanent wide cracks	0.10-0.20	>10
C, CL, SC, sCL	Very fine or fine prismatic, angular blocky or platy	0.01-0.03	0.01-0.1
C, SC, sC, CL, sCL, SL, S, sCL	Fine and medium prismatic, angular blocky and platy	0.03-0.08	0.1-0.4
Light CL, S, SL, very fine sL, L	Medium prismatic and subangular blocky	0.06-0.12	0.3-1.0
Fine sandy loam, sandy loam	Coarse subangular block and granular, fine crumb	0.12-0.18	1.0-3.0
Loamy sand	Medium crumb	0.15-0.22	1.6-6.0
Fine sand	Single grain	0.15-0.22	1.6-6.0
Medium sand	Single grain	0.22-0.26	>6
Coarse sand and gravel	Single grain	0.26-0.35	>6

Source: Anonymous[2]

Wetted width, $w = K_2\left(V_w\right)^{0.22}\left(\frac{C_s}{q}\right)^{-0.17}$

$$= 0.031(14.4)^{0.22}\left(\frac{0.25}{\frac{24x3600}{2}}\right)^{-0.17}$$

$$= 0.031x1.798x9.834$$
$$= 0.55m$$

For a pair of drippers at each plant and 14.4l water volume per application,

Depth of wetting front, $z = K_1 (V_w)^{0.63} \left(\frac{C_s}{q} \right)^{0.45}$

$$= 29.2(14.4/2)^{0.63} \left(\frac{0.25}{\frac{24x3600}{2}} \right)^{0.45}$$

$$= 29.2x3.468x2.356x10^{-3}$$

$$= 0.24m$$

Wetted width, $w = K_2 \left(V_w\right)^{0.22} \left(\frac{C_s}{q} \right)^{-0.17}$

$$= 0.031(14.4/2)^{0.22} \left(\frac{0.25}{\frac{24 \times 3600}{2}} \right)^{-0.17}$$

$$= 0.031 \times 1.54 \times 9.834$$

$$= 0.47m$$

From the above trials it appears that using single dripper of capacity 2l/h gives much higher depth and pair of drippers gives less depth of penetration to the suggested depth of 30cm.

Let 1l/h drippers to be used in pair

Therefore, $z = K_1 (V_w)^{0.63} \left(\frac{C_s}{q} \right)^{0.45}$

$$= 29.2(14.4/2)^{0.63} \left(\frac{0.25}{\frac{24 \times 3600}{1}} \right)^{0.45}$$

$$= 29.2 \times 3.468 \times 3.2184 \times 10^{-3}$$

$$= 0.326 \cong 0.33m$$

$$w = K_2 \left(V_w\right)^{0.22} \left(\frac{C_s}{q} \right)^{-0.17}$$

$$= 0.031(14.4/2)^{0.22}\left(\frac{0.25}{\frac{24\times3600}{1}}\right)^{-0.17}$$

$$= 0.031\times1.54\times8.7409$$

$$= 0.416 \cong 0.42m$$

The depth of penetration 33cm is quite good and may be accepted.

Since two drippers are used the wetted area around the plant bottom for 14.4l volume of application, $A_w = 2\times\pi\frac{d^2}{4} = 2\times\pi\frac{(0.42)^2}{4} = 0.277m^2$

Wetted volume of soil, Vs= 0.277m x0.33m=0.09m^3

Similarly, the depth, width and wetted area are calculated for different irrigations and tabulated in Table 4.11.

Table 4.11: Depth, width and volume of wetting at different application of water in different months

Months	Water requirement, l/plant/ irrigation,	Depth of penetration (z), m	Width of wetting (w), m	Volume of wetted soil (Vs), m^3	Water content in soil before irrigation, m^3/m^3
(1)	(2)	(3)	(4)	(5)	(6)
May	2.88	0.12	0.29	0.016	0.206
June	2.88	0.12	0.29	0.016	0.206
July	5.7	0.19	0.34	0.034	0.218
Aug	8.1	0.23	0.37	0.049	0.220
Sept	14.4	0.33	0.42	0.091	0.227
Oct	16.8	0.36	0.43	0.105	0.226
Nov	25.0	0.50	0.48	0.181	0.247
Dec	22.0	0.43	0.46	0.143	0.232
Jan	19.25	0.39	0.45	0.122	0.228
Feb	22.0	0.43	0.46	0.142	0.231
Mar	17.6	0.37	0.44	0.111	0.227
April	20.0	0.40	0.45	0.127	0.228
May	20.0	0.40	0.45	0.127	0.228

Calculation of water content in soil at plant bottom before irrigation:

Let the month of June is taken into consideration.

Field capacity of soil =38.56% on volumetric basis (Anonymous)[2].

Volume of water in wetted soil volume on irrigation =0.016x0.3856=0.0062m^3

Water volume in wetted soil at the time of irrigation=0.0062-0.00288=0.0033m^3

Volumetric moisture content at the time of irrigation=0.0033/0.016=0.206m/m

Soil:

This suction of soil is calculated by using the equation,

$$\theta = 0.245e^{-0.41\log_{10}\psi} \tag{4.38}$$

This equation has been developed for the present soil by using the data as obtained from Anonymous[2].

For the month of May, $\theta = 0.245e^{-0.41\log_{10}\psi}$

$$0.206 = 0.245e^{-0.41\log_{10}\psi}$$

$$0.841 = e^{-0.41\log_{10}\psi}$$

$$-0.173 = -0.41\log_{10}\psi$$

$$0.422 = \log_{10}\psi$$

$$\psi = 2.647$$

Similarly the soil suction in different months are calculated and tabulated in column 7 of Table 4.12.

Table 4.12: Wetting and suction of soil in different months11

Months	**Water requirement, l/ plant/irrigation**	**Depth of penetration (z), m**	**Width of wetting (w), m**	**Volume of wetted soil (V_s), m^3**	**Water content in soil before irrigation, m^3/m^3**	**Soil suction at the time of irrigation, bar**	**Permissible soil suction, bar**
(1)	(2)	(3)	(4)	(5)	(6)	(7)	(8)
May	2.88	0.12	0.29	0.016	0.206	2.65	0.40
June	2.88	0.12	0.29	0.016	0.206	2.65	0.40
July	5.7	0.19	0.34	0.032	0.218	1.93	0.65
Aug	8.1	0.23	0.37	0.049	0.220	1.83	0.90
Sept	14.4	0.33	0.42	0.091	0.227	1.54	1.40
Oct	16.8	0.36	0.43	0.105	0.226	1.57	1.70
Nov	25.0	0.50	0.48	0.181	0.247	0.96	2.75
Dec	22.0	0.43	0.46	0.143	0.232	1.36	3.50
Jan	19.25	0.39	0.45	0.122	0.228	1.50	4.00
Feb	22.0	0.43	0.46	0.142	0.231	1.39	3.50
Mar	17.6	0.37	0.44	0.111	0.227	1.54	1.40
April	20.0	0.40	0.45	0.127	0.228	1.50	0.40
May	20.0	0.40	0.45	0.127	0.228	1.50	0.40

In examining the Table 4.12 it appears that soil moisture suctions have exceeded the permissible limits during March-September at fixed proposed intervals. It is therefore necessary to apply more water during these months to maintain the suctions within the permissible limits. In consideration to the permissible limits of suctions (Table 3.7) and by using Eq. 4.38 $[\theta = 0.245e^{-0.41\log_{10}\psi}]$ the modified moisture content at the time of irrigations, water application requirements, etc. are calculated and tabulated in Table 4.13.

Month: May & June

Permissible soil moisture suction: 0.4bar

Soil moisture needed at the time of irrigation,

$$\theta = 0.245e^{-0.41\log_{10}\psi}$$

$$= 0.245e^{-0.41\log_{10}(0.4)}$$

$$= 0.288\text{m/m}$$

Deficit of soil moisture at the time of irrigation due to less application of water = 0.288-0.206 = 0.082m/m.

Application of 2.88liter water brings the soil moisture content 0.206m/m to 0.3856m/m (field capacity) to 0.016m3 soil. The increase in soil moisture content = 0.3856-0.206 =0.1796m/m. It is, therefore, the additional water necessary to increase the 0.082m/m water to the same volume of soil

$$= \frac{2.88l}{0.1796} x0.082 = 1.315l$$

Modified application rate = 2.88+1.315 =4.195≈4.2liter

Using Schwarzman & Zur (1985) equation,

Depth of wetting,

$$z = K_1(V_w)^{0.63}\left(\frac{C_s}{q}\right)^{0.45}$$

$$= 29.2(4.2/2)^{0.63}\left(\frac{0.25}{\frac{24x3600}{1}}\right)^{0.45}$$

$$= 29.2x1.596x3.2184x10^{-3}$$

$$= 0.15m$$

Width of wetting,

$$w = K_2 \left(V_w\right)^{0.22} \left(\frac{C_s}{q}\right)^{-0.17}$$

$$= 0.031\left(4.2/2\right)^{0.22} \left(\frac{\frac{0.25}{24 \times 3600}}{1}\right)^{-0.17}$$

$= 0.031 \times 1.177 \times 8.7409$

$=0.319$m

Volume of wetted area $= \pi w^2 / 4 \times zx^2 = \pi(0.319)^2 / 4 \times 0.15x^2 = 0.024m^3$

Volume of water in wetted soil volume = 0.024 × 0.3856 = 9.25 × $10^{-3}m^3$

Volume of water in soil at the time of irrigation

=9.25 × $10^{-3}m^3$-2.88liter

=9.25 × 10^{-3}-0.00288

=6.37 × $10^{-3}m^3$

Volumetric moisture content = 6.37 × 10^{-3}/0.024 = 0.265m/m

The moisture content 0.265m/m is less than the required moisture content 0.288m/m. Let the application of water is further increased by 15%.

So, modified application rate = 4.2 × 1.15 = 4.83liter

With the application of 4.83l and calculating by using Schwarzman & Zur (19 85)

z = 0.164m

w = 0.329m

Wetted volume of soil = 0.0279m^3

Volume of water in wetted soil volume = 0.0279 × 0.3856 = 0.011m^3

Volume of water at the time of irrigation = 0.011-0.00288 = 8.12 × $10^{-3}m^3$

Soil moisture content at the time of irrigation = 8.12 × 10^{-3}/0.0279 = 0.29m/m

0.29m/m > 0.288m/m. This is accepted.

$$\theta = 0.245e^{-0.41\log_{10}\psi}$$

$$or, \frac{0.29}{0.245} = e^{-0.41\log_{10}\psi}$$

$$\therefore -0.41 = \log_{10}\psi$$

$$\psi = 0.39bar$$

Similarly the deficits of soil moisture contents at the time of water application are calculated for the other months. Depending on the difference of soil moisture suctions previously estimated and the permissible soil moisture suctions (Table 4.13), in modified application of water 5 to 15% additional water were required to be applied in excess to the calculated amount of apparent deficit of moisture at the time of water application. The details are shown in Table 4.13.

Drip network design

Main

The pump is at the middle of the field. Therefore, the flow gets divided equally in two opposite direction.

Rate of application =2l/h/plant

Rate of discharge in a side = 2x4540/2= 4540l/h = 1.26l/s

Rate of pumping = 1.26x2 = 2.52l/s.

Let us assume 6m pieces of main pipe and 0.5m equivalent length for friction of each joint.

No. of joints on each side= (170-2)/(2x6) = 14

Equivalent length due to joint = 14x0.5 =7.0m

Let friction loss equivalent due to bends & other fittings = 5.0m

Total main length = 168/2+7+5 = 96m

Let the diameter of the main = 50mm

Using Hazen-William formula,

Months	Water requirement, l/plant/ irrigation,	Apparent deficit of the soil moisture, l/plant/day	% increase to modified water requirement	Modified water requirement, liter	Depth of penetration (z), m	Width of wetting (w), m	Volume of wetted soil (Vs), m3	Water content in soil at the time of irrigation, m3/m3	Soil moisture suction at the time of irrigation, bar	Permissible soil moisture suction, bar
(1)	(2)	(3)	(4)	(5)	(6)	(7)	(8)	(9)	(10)	(11)
May	2.88	1.315	15	4.83	0.164	0.329	0.0279	0.29	0.39	0.40
June	2.88	1.315	15	4.83	0.164	0.329	0.0279	0.29	0.39	0.40
July	5.7	2.50	10	8.67	0.237	0.373	0.0518	0.274	0.53	0.65
Aug	8.1	1.42	5	10.0	0.259	0.386	0.0606	0.252	0.85	0.90
Sept	14.4	0.038	5	15.16	0.337	0.423	0.0948	0.232	1.36	1.40
Oct	16.8	-	-	16.8	0.36	0.43	0.105	0.226	1.57	1.70
Nov	28.0	-	-	28.0	0.50	0.48	0.181	0.231	1.39	2.75
Dec	22.0	-	-	22.0	0.43	0.46	0.143	0.232	1.36	3.50
Jan	19.25	-	-	19.25	0.39	0.45	0.122	0.228	1.50	4.00
Feb	22.0		-	22.0	0.43	0.46	0.142	0.231	1.39	3.50
Mar	17.6		10	19.41	0.393	0.447	0.1233	0.243	1.05	1.40
April	20.0		15	31.75	0.536	0.498	0.2088	0.29	0.39	0.40

Head loss, $h_l = 3.98x10^5 x \frac{q^{1.852}}{d^{4.871}} xL$

$$= 3.98x10^5 x \frac{1.26^{1.852}}{50^{4.871}} x96$$

$$= 3.98x10^5 x \frac{1.534}{1.889x10^8} x96$$

$$= 0.31\text{m}$$

The loss of head in the main is less than 1m. This is accepted.

Lateral

No. of plants/lateral = 55/2 =27.5 =28(say)

No. of dripper on a lateral = 28x2 = 56

Equivalent length for friction loss due to drippers = 56x0.5 = 28m

Total equivalent length of a lateral = 55 +28 + 3 (field road) =86m

Discharge of a lateral = 56x1 =56l/h= 0.0155l/s

Let diameter of the lateral = 10mm

Head loss in lateral, $h_l = 3.98x10^5 x \frac{q^{1.852}}{d^{4.871}} x(L + L_e)$

$$= 3.98x10^5 x \frac{0.0155^{1.852}}{10^{4.871}} x85$$

$$= 3.98x10^5 x \frac{4.45x10^{-4}}{7.43x10^4} x85$$

$$= 0.20\text{m}$$

10mm lateral pipe is accepted.

Pump selection

Static head = 15m (similar to Example 4.13)

Friction head loss in main = 0.31m

Friction loss in lateral = 0.20m

Minimum average head required on dripper = 10m

Friction loss in control head = 10m

Total loss of head in drip unit = 20.51

Total head = 15 + 20.51 =35.51m

Horse power of the pump, $HP = \frac{QxH}{75xe}$

$$= \frac{2.52x35.51}{75x0.6}$$

$$= 1.99$$

It is suggested to purchase 2.5hp capacity pumpset.

Assuming the area under different crop cultivation

In this design we may have the provision of irrigating all the plots at a time and separately to any plot. Therefore, in each plot the sub mains are proposed as shown in Fig. 4.27. The main line extends east-west wise up to the B & D plots from where sub mains could be run through the sides of the plots and connected to the main at minimum length of the main.

1. Water requirement

No. of plants in orchard

$$= \frac{55m \times 40m}{5m \times 5m} + \frac{(55m \times 60m - (5m \times 5m)}{5m \times 5m} = 88 = 131 = 219$$

No. of plants in papaya

$$= \frac{55m \times 55m - 15m \times 10m}{2m \times 2m} + \frac{55m \times 70m}{2m \times 2m} = 718.75 + 962.5 = 1681.24 \cong 1680$$

No. of banana plants

$$= \frac{55m \times 40m}{2m \times 2m} + \frac{55m \times 30m}{2m \times 2m} + \frac{550m \times 40m}{2m \times 2m}$$

$$= 550 + 412.5 + 550 = 1512.5 \cong 1512$$

Water requirement in orchard

$= E_{pan} \times K_p \times K_c \times A_w \times S_p$

=6.5mm × 0.8 ×1.0 × 0.3 × 5m × 5m

= 39liter/day/plant

= 39 × 219 liter/day

= 8541liter/day

Water requirement in papaya

$= E_{pan} \times K_p \times K_c \times A_w \times S_p$

=6.5mm × 0.8 × 1.0 × 0.5 × 2m x 2m

= 10.4liter/day/plant

= 10.4 × 1680liter/day

= 17472l/day

Water requirement in banana

$= E_{pan} \times K_p \times K_c \times A_w \times S_p$

= 11.44 liter/day/plant

= 11.4 ×1512 liter/day

= 17297.28 liter/day

Total water requirement = 8541 + 17472 + 17297 = 43310.28 l/day

2. Selection of drippers

Similar to Example 4.13 we may use 4l/h drippers 2 in numbers/plant for orchard and 1l/h 2 in numbers for each plant in banana and papaya.

3. Selection of main

If all the plots are irrigated simultaneously, the rate of application is as below.

As shown in Fig. 4.27 the required main line length is 90m of which 57m on the right and 33m on the left side of the tube well.

a. On 57m main side = Plot A (550 plants × 2l/h/plant)
+ Plot B (550 plants × 2l/h/plant)
+ Plot C (131 plants × 8l/h/plant)
+ Plot G (718 plants × 2l/h/plant)
= 1100 + 1100 + 1048 + 1436 = 4684l/h = 1.30l/s

b. On 33m main side = Plot D(962 plants x 2l/h/plant)
+ Plot E (88 plants × 8l/h/plant)
+ Plot F (412 plants × 2l/h/plant)
= 1924 + 704 +824 =3452 l/h = 0.96l/s

Total discharge = 2.26l/s

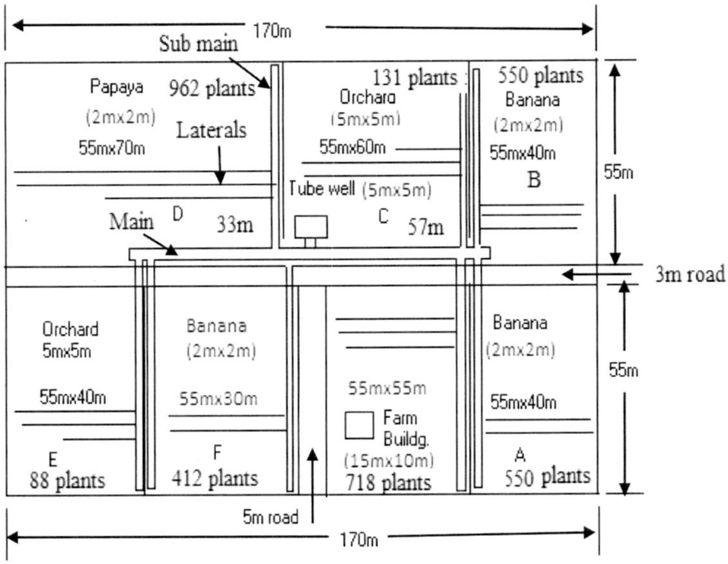

Figure 4.27: Layout of field for various crops

Let the diameter of main = 75mm

Friction loss in the main at the higher discharge side (57m),

$$h_l = 3.98 \times 10^5 \times \frac{q^{1.852}}{d^{4.871}} \times (L + L_e)$$

$$= 3.98 \times 10^5 \times \frac{(1.30)^{1.852}}{(75)^{4.871}} \times (57 + 5)$$

$$= 3.98 \times 10^5 \times \frac{(1.30)^{1.852}}{(63)^{4.871}} \times (57 + 5)$$

$$= 3.98 \times 10^5 \times \frac{1.626}{5.816x10^8} \times 62$$

$$= 0.07m$$

The friction loss is much less. Low diameter main may be tested. Let the diameter of main to be 50mm. Therefore,

$$h_l = 3.98 \times 10^5 \times \frac{(1.30)^{1.852}}{(50)^{4.871}} \times 62$$

$$= 3.98 \times 10^5 \times \frac{1.626}{1.89 \times 10^8} \times 62$$

$$= 0.21m$$

Selection of sub mains

After examining the discharges it appears that maximum discharge 1436l/h or 0.4l/s occurs in plot G. Let the sub main size be 30 mm.

The friction loss, $h_l = 3.98\times10^5 \times \frac{q^{1.852}}{d^{4.871}}\times(L+L_e)$

$$= 3.98\times10^5 \times \frac{(0.4)^{1.852}}{(30)^{4.871}}\times(55+5)$$

$$= 3.98\times10^5 \times \frac{0.18}{1.57x10^7}\times 60$$

$$= 0.27m$$

The head loss 0.27 is acceptable. Thus, the sub main to be 30mm

Selection of laterals

No. of plants in a lateral = 120/(2 × 5) = 12

No. of drippers in a lateral = 12 × 2 = 24

The discharge of a lateral = 24 × 4 x 2 = 96l/h – 0.027l/s

Length of the lateral = 120/2-5/2 = 0-2.5 = 57.5m

Let us assume the barb (local) loss due to penetration of drippers in lateral is equivalent to 0.2m [Biswas *et al.* (2006) & Reddy *et al.* (2001)]

Equivalent length for barb loss in a lateral, Le = 24 × 0.2 = 4.8m

Let diameter of lateral pipe = 10mm

Head loss in a lateral, $h_l = 3.98x10^5 \times \frac{q^{1.852}}{d^{4.871}}\times(L+L_e)$

$$= 3.98\times10^5 \times \frac{(0.027)^{1.852}}{10^{4.871}}\times(57.5+4.8)$$

$$= 3.98\times10^5 \times \frac{1.24\times10^{-3}}{7.43\times10^4}\times 62.3$$

$$= 0.33m$$

The head loss of 0.33m in lateral is acceptable. Therefore, 10mm laterals are selected.

Selection of pump

Total operating head of pump = Static head (suction head + delivery head) + average operating head of drip system + friction loss in main + friction loss in sub main + friction loss in lateral + head loss of equipments in control head + head loss or gain due to elevation.

Static head = 15m (say)

Average operating head of drip system = 10m

Friction loss in main = 0.21m

Friction loss in sub main = 0.27m

Friction loss in lateral = 0.33m

Head loss in control head = 10m (say)

Head gain or loss = nil

Total head = 35.81m

Horse power of the pump, $HP = \dfrac{Q \times H}{75 \times e}$

$$= \frac{2.26 \times 35.81}{75 \times 0.6}$$

$$= 1.8$$

It is found that the capacity of the pump set is almost same either all the plots are cultivated with banana crop or by different crops and provisions are made to irrigate any plot independently. Thus, the previously selected pump capacity of 2.5hp is accepted and that to be in the safe side.

4.0 Questions & Problems

4.1. State the methods of determining pipe sizes in laterals.

4.2. Discuss the Darcy-Weisbach, Hazen-William and Scobey equations for head losses in pipes.

4.3. The plastic pipe (C=150) used tras a drip lateral of diameter 12mm having 50 distributors of capacity 10l/h and spaced at 5m interval. What is the friction head loss in pipe?

Ans. 5.84m

4.4. What is the expected local head loss in a lateral pipe of inside diameter 9.84mm with in-line 50 numbers of distributors of capacity 5l/h?

Ans. 0.55m

4.5. What is the tapered pipe lateral/ how the head loss of this lateral is measured?

4.6. Derive the expression for hydraulic head $H_i = H - R_i \Delta H \pm R_i' \Delta H'$ in a drip lateral for any length ratio.

4.7. A orchard field of spacing 5mx5m uses two numbers of 4lps dripper to each plant. The drippers operate at 11m pressure head at the head end of a 150m lateral and 12mm diameter. The laterals are laid in a up slope of 0.5%. The flow characteristics of drippers is $q = K_d H^{0.5}$. Determine the pressures at length ratios of 0.1, 0.2,....1.0 and dripper flow variation in the lateral.

Ans. Length ratios 0.1, 0.2, ...1.0 = 10.35m, 9.81m, 9.00m, 8.72m, 8.50m, 8.34m, 8.21m, 8.12m, 8.04m. Hvar=0.18, qvar=0.094.

4.8. State the assumptions of head losses in a drip lateral pipe following Keller & Karmeli.

4.9. Determine the uniformity of emission in a flat field for the following.

Lateral length = 150m

Distributor discharge = 5lps

Distributor characteristics = $q = 0.65H^{0.85}$

Average pressure = 10m

Manufacturer's coefficient of variation, CV = 0.03%

Spacing of distributors = 4.5mx4.5m

PVC pipes available in the market = ID(mm)-9.55mm, 11.50mm, 13.55mm, 15.45mm

Ans. 94.53%

4.10. Select the appropriate answer from the following.

1. Friction coefficient f in a laminar flow is

a) 16/Re b) 32/Re

c) 64/Re d) 128/Re

2. The generalized equation for head loss in pipe may be stated as

a) $K\frac{lq^{m+1}}{d^{2m+n}}$ b) $K\frac{lq^{m}}{d^{2m+n}}$

c) $K\frac{lq^{m+1}}{d^{3m+n}}$ d) $K\frac{lq^{m+1}}{d^{m+n}}$

3. The same flow of 0.1lps occurs through 15mm & 10mm plastic lateral pipe. How much times the friction head loss in 10mm pipe compared to 15mm?

a) 1.5 b) 4.8

c) 5.6 d) 7.2

4. A 100m lateral pipe is run 2% upslope for 30m and the rest 1.5% down slope. The energy gain due to the slope is

a) -0.6m b) -0.05
c) 0.05 d) 0.45

5. Keller & Karmeli proposed average head in a lateral at length ratio

a) 0.21 b) 0.39
c) 0.49 d) 0.77

6. The average pressure and pressure at head end in a lateral are 8.5m and 10m respectively. The pressure at the lateral tail end may be

a) 8.05m b) 8.15m
c) 8.25m d) 8.45m

7. In a PVC pipe of 100m length and diameter 16mm discharges 300l/h. The head loss is about

a) 0.75m b) 1.25m
c) 1.53m d) 1.63m

8. For a certain diameter PVC pipe head loss for 100m pipe is presented by $\Delta H = 493.22q^a$.

If the q(discharge) in the pipe is 0.1l/s and head loss 6.934m, the value of a is

a) 0.851 b) 1.852
c) 1.895 d) 2.875

9. The difference between the energy drop ratio following the Darcy-Weisbach and Hazen-William method at 0.5 length ratio is about

a) 8.75×10-2 b) 1.35×10-2
c) 8.61×10-2 d) 1.70×10-1

10. A drip main line runs 3 steps each of 50m length at 2% down slope, 3% up slope and 1% down slope respectively. The energy gain is

a) 0.0m b) -0.5m
c) +0.5m d) +1.0m

11. The friction energy loss in a section of drip line is 10%. The section is laid in 2% up slope. The total energy slope is

a) -12% b) -8%
c) +8% d) +12%

12. At length ratio 1 the position of a 100m pipe is

    a) 0.0m b) 25m

    c) 75m d) 100m

    Ans. 1. (c) 2. (b) 3. (d) 4. (d) 5. (b) 6. (a) 7. (c)8. (b) 9. (b) 10. (a) 11. (a) 12. (d)

4.11. Write True or false of the following.

1. Energy loss in drip lateral increases as it goes downward.
2. Sub-main of drip system is suggested to follow the steepest slope as far as practicable.
3. For better distribution of water at less energy loss the pump may be installed at the middle of the field.
4. Maximum energy loss in a lateral is 10% of average pressure.
5. The maximum energy loss in a submain will be at the middle section whose magnitude is equal to 0.36 ΔH
6. The magnitude of multiplication factor (F) for friction loss increases with the numbers of emitters in drip lateral pipe.

Ans. 1. False 2. True 3. True 4. True 5. True 6. False

References

Biswas, R.K., De, P., Das,J., Poddas, S. & Sarkar, A. (2005). Drip irrigation system design-an analytical approach. 8th International symposium on water crisis, global warming and sea level rise in Bengal Basin held on Dec 16-18, 2005 at BCKV, West Bengal, India.

FAO(1980). Vermeiren, I and Jobling, G.A. Irrigation and Drainage Paper 36, FAO. Rome.

Keller, J. and Karmeli, D. (1974). Trickle Irrigation Design Parameters. Trans. ASAE.

Michael, A.M. (1978). Irrigation Theory aand Practices. Vikas Publishing House Pvt. Ltd., New Delhi.

Reddy, K.Y., Satyanarayana, T.V., Appa Rao, D., Sathya Prasad, A. & Madhava, Ml (2001. Evaluation of On-line Trickle Irrigation Emitter Barb Losses. Micro Irrigation. Central Board of Irrigation & Power. Publication No.282.

Wu, I and Gitlin, H.M. (1974). Drip irrigation design based on uniformity. Trans. ASAE. Vol.17. No.3.

5

Distributors

Introduction

There is different type of distributors available in the market. Depending on the basic principle of working they may be broadly classified into three major groups as below (FAO, 1980).

i. Distributors in which head loss takes place through small diameter tubes
ii. Distributors in which head loss takes place through some orifice control
iii. Distributors in which head loss takes place through a vortex action

There are many adaptations to distributors though the working distributors fall under the above stated basic principle. Tiny perforation in the laterals or porous materials allows leaking out the water serves the purpose of distributors. The compensating type distributors have long path in it to dissipate the energy and drips almost same rate within certain range of pressure variation. Whatever may be the distributors proposed to be used, the discharge characteristics should be known. The discharge characteristics should be provided by the manufacturers and preferably certified by any authorized body; otherwise to be tested by the designer before recommendation to large-scale use.

Types of Distributors

Small diameter tube or long path distributors

Micro tubes

The micro tubes are the black polythene pipe approximately 0.5 to 1.5mm internal diameter (Fig. 5.1). The discharge through these pipes may be expressed by the following equation (FAO, 1980),

$$q = al^b H^c D^d \tag{5.1}$$

D (mm) Coefficient	0.5	0.6	0.7	0.8	0.9	01.0	1.1
a	0.86	0.91	1.02	1.14	1.16	1.28	1.38
b	-0.78	-0.75	-0.72	-0.68	-0.65	-0.62	-0.58
c	0.85	0.82	0.78	0.72	0.72	0.60	0.65
d	3.1	3.1	3.1	3.1	3.1	3.1	3.1

Courtesy: FAO, 1980

where, q = the discharge of the micro tubes, l/h

l = the length of the micro tube, m

H = the operating pressure, m of water

D = the internal diameter of the micro tube, mm

a, b, c, d = the coefficients depending on the value of D (Table 5.2).

The pressure in a lateral decreases towards the tail end and therefore the discharges to be more at the head end and progressively less towards the tail end. To have good uniformity of flow through different distribution points the lengths of the micro tubes may be calculated following Eq. 5.1 with known value of pressure distribution throughout the length of the lateral. It is possible to use the micro tubes for consecutive seasons if the crop spacing is unaltered. Difficulties in using micro tubes reported to be its cumbersome process of design and application to field, hindrance to movement through the high concentration of micro tubes in close spacing crops and chances of damage by the rodents.

Example 5.1 Determine the length of micro tubes of diameter 0.5mm at 20, 50 & 80m length of a lateral for discharging at a rate of 0.5l/h with pressure gradient of 5%. The initial pressure at the lateral is 11m.

Solution

Initial pressure head of the lateral = 11m

So, pressure at 20m length of lateral = $11 - \frac{20 \times 5}{100} = 10m$

pressure at 50m length of lateral = $11 - \frac{50 \times 5}{100} = 8.5m$

Fig. 5.1 Micro tubing

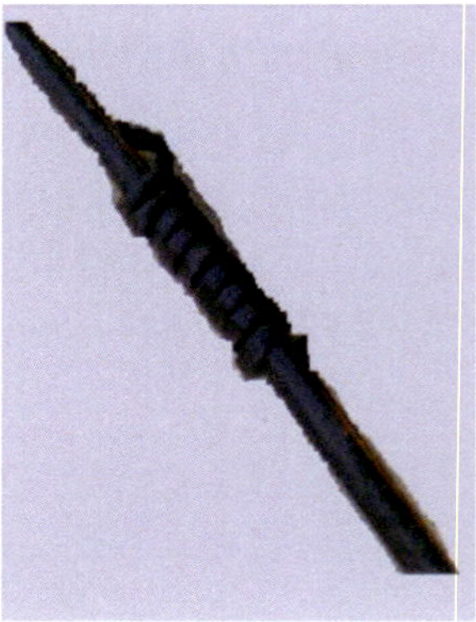

Fig. 5.2 Pre-coiled microtubes

pressure at 80m length of lateral $= 11 - \frac{80 \times 5}{100} = 7.0m$

$q = al^b H^c D^d$

where, $q = 0.5l / h$

$H = 11m$

a, b, c, d are constants and whose values are 0.86, -0.78,0.85 and 3.1 respectively as obtained from Table 5.1.

At 20m length of lateral

$0.5 = 0.86 \times l^{-0.78} \times 10^{0.85} \times 0.5^{3.1}$

or, $0.5 = 0.86 \times l^{-0.78} \times 7.073 \times 0.1167$

$$\therefore l^{-0.78} = \frac{0.5}{0.86 \times 7.073 \times 0.1167} = \frac{0.5}{0.704} = 0.71$$

$$\therefore l = (0.71)^{-\frac{1}{0.78}} = 1.55m$$

At 50m length of lateral

$0.5 = 0.86 \times l^{-0.78} \times 8.5^{0.85} \times 0.5^{3.1}$

$$\therefore l^{-0.78} = \frac{0.5}{0.86 \times 6.166 \times 0.1167} = \frac{0.5}{0.614} = 0.80$$

$$\therefore l = 1.33m$$

At 80m length of lateral

$$0.5 = 0.86 \times l^{-0.78} \times 7.03^{0.85} \times 0.5^{3.1}$$

$$\therefore l^{-0.78} = \frac{0.5}{0.86 \times 5.228 \times 0.1167} = \frac{0.5}{0.5256} = 0.95$$

$$\therefore l = 1.067m$$

Example 5.2 Determine the length of micro tube for 0.5, 0.6, 0.7 & 0.8mm diameter at 10m interval in a lateral of 100m length and diameter 10mm. Assume rate of flow 0.05l/s, initial pressure head 15m and micro tubes are spaced 1.0m.

Solution

$$q = 0.05l / s, d = 0.5mm$$

$$\therefore v = \frac{0.05 / 1000}{\frac{\pi (10 / 1000)^2}{4}} = 0.63 = 63cm / s, C = 150$$

Using Hazen-William equation (Eq. 4.2) for friction loss in lateral pipe,

$$h_l = \frac{3.022 l v^{1.852}}{C^{1.852} d^{1.167}}$$

$$= \frac{3.022 \times 100 \times 63^{1.852}}{150^{1.852} \times 10^{1.167}}$$

$$= \frac{649640.59}{157442} = 4.126m$$

Using Hazen-William equation (Eq. 4.14) for energy gradient in lateral,

$$R_i = 1 - (1-i)^{2.852}$$

where, R_i = energy drop ratio

i = length ratio l/L,

l = length of the lateral measured from head end of the lateral

L = total length of lateral.

At 10m intervals, the length ratios (i_s) are 0.1,0.2,0.3,0.4,0.5,0.6,0.7,0.8,0.9 & 1.0.

When, $i = 0.1, R_i = 1 - (1-0.1)^{2.852} = 0.2535$

So, head at 10m length (i = 0.1) = head at head end-head lost up to 0.1 length ratio

$= 15 - 4.125 \times 0.2535$

$= 13.93m$

Similarly heads at other length ratios are calculated and tabulated as below.

Length ratio	0.1	0.2	0.3	0.4	0.5	0.6	0.7	0.8	0.9	1.0
Head, m	13.93	13.06	12.36	11.83	11.45	11.17	11.01	10.92	10.88	10.87

Spacing of micro tubes = 1.0m

No. of micro tubes in lateral = 100

Discharge through each micro tubes = $\frac{0.05}{100} = 0.0005 = 1.8l / h$

We have, $q = a l^b H^c D^d$

$$\therefore 1.8 = 0.86 l^{-0.78} 13.93^{0.85} 0.5^{3.1}$$

i. $\therefore 1.8 = 0.86 \times l^{-0.78} \times 13.93^{0.85} \times 0.5^{3.1}$ at length ratio(i)=0.1 & dia. of micro tube = 0.5mm

$$\therefore l^{-0.780} = \frac{1.8}{0.86 \times 9.38 \times 0.1167} = 1.92$$

or, $l = 0.433m$

ii. Similarly, $1.8 = 0.91 \times l^{-0.72} \times 13.93^{0.82} \times 0.6^{3.1}$, at I =0.1 & dia. of micro tube = 0.6mm

$$\therefore l^{-0.75} = \frac{1.8}{0.91 \times 8.67 \times 0.205} = 1.106$$

or, $l = 0.86m$

iii. Similarly, $1.8 = 1.02 \times l^{-0.72} \times 13.93^{0.72} \times 0.7^{3.1}$, at I =0.1 & dia. of micro tube = 0.7mm

$$\therefore l^{-0.72} = \frac{1.8}{1.02 \times 7.0 \times 0.33} = 0.69$$

or, $l = 1.67m$

iv. Similarly, $1.8 = 1.14 \times l^{-0.68} \times 13.93^{0.75} \times 0.8^{3.1}$, at I =0.1 & dia. of micro tube = 0.8mm

$$\therefore l^{-0.68} = \frac{1.8}{1.14 \times 7.21 \times 0.5} = 0.437$$

or, $l = 3.37m$

Similarly the length of the micro tubes of other diameters have been calculated and tabulated as below.

Dia. (mm)	Length of micro tubes at different length ratios, m									
	0.1	**0.2**	**0.3**	**0.4**	**0.5**	**0.6**	**0.7**	**0.8**	**0.9**	**1.0**
0.5	0.433	0.4	0.38	0.36	0.34	0.34	0.33	0.33	0.33	0.33
0.6	0.86	0.81	0.76	0.73	0.70	0.67	0.67	0.67	0.66	0.66
0.7	1.67	1.57	1.48	1.43	1.34	1.33	1.31	1.31	1.24	1.24
0.8	3.37	3.20	3.00	2.81	2.77	2.65	2.60	2.57	2.56	2.56

Long flow path integrated distributors

The use of micro tubes has the disadvantage of getting displaced inadvertently. To the aim at overcome this problem the micro tubes are sometime made to fit as a part of the lateral and others are suitably fitted on the side of the lateral.

i. Pre-coiled micro tube

These are the distributors made of micro tubes coiled around the lateral with one end inserted in to the pressure fit hole in the lateral (Fig. 5.2).

ii. Internal spiral / labyrinth distributors

These are in principle similar to micro tubes. The narrow passage in molded plastic accessories serves the purpose of micro tubes to cause the energy loss and slow discharge of water (Fig. 5.3-5.7). The molded distributors can be easily a part of the lateral pipe without protruding and dangling. It can be set and withdraw at any point on the lateral at any time without causing much injury to lateral. This helps to use the distributors for subsequent season and at variable spacing in laterals. This distributor may be of multiple type favoring more rate of discharge sometime requires in light soils.

These distributors may be cleaned by flushing action or by dismantling the body and cleaning the spiral passage. Self-cleaning distributors overcome the tendency of clogging to tiny passage by flushing at the beginning and at the end of each watering. In self-flushing, at low pressure the water flows freely and flushing the passage, as the pressure increases a ball or a spring closes the orifice at the appropriate stage. The elastomer ball (disc) seated against the spiral path on the plastic disc. As the pressure increases in the inlet, the elastomer disc deforms in to the spiral path reduces the cross-section of flow. Thus, higher head loss occurs at high pressure head at the inlet. When the pressure is low in the inlet at the beginning and end of the flushing, the elastomer ball seats freely on the spiral groove causing some flow in the passage, which allows flushing action. Distributors having other principle of operation are also available in the market. The details of which may be obtained from the manufacturers. The distributors which provide almost rated discharge at wide pressure range are called compensating distributors. Since the discharge of the distributors becomes independent to head of pressure, the exponent x in discharge-head relation is zero.

Orifice distributors

Perforated single chamber tube

The simplest form of orifice distributor is the perforated single chamber tube (or lateral). The perforations are made at regular interval with small diameters. It is very difficult to have equal size diameters through drilling or punching leading to poor uniformity in discharges through the holes. Due to variation in pressure along the lateral and thereby the discharges the length of lateral should not be more than 60m.

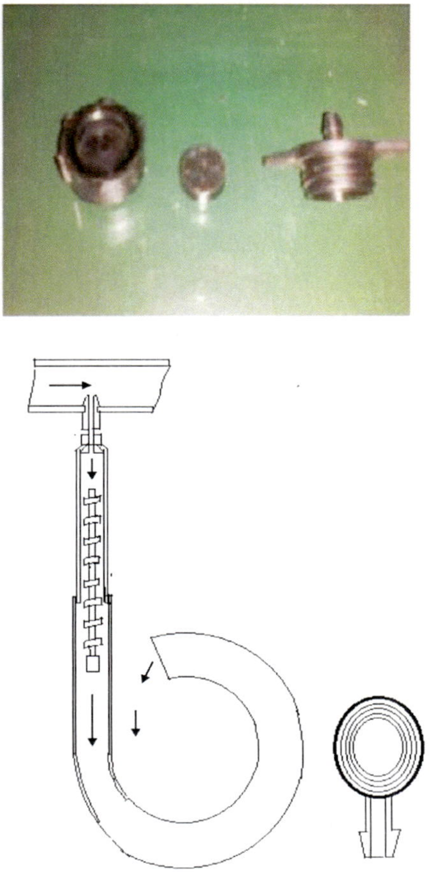

Figure 5.3: Inside of an internal spiral moded distributor

Calibrated orifice

To overcome the problem of poor application efficiency due to improper diameters and control in perforated laterals, calibrated orifice distributors are made inserted and have fixed geometry (Fig. 5.6). The incoming water jet through the orifice is broken by a baffling action allowing the water to drips. The discharge through these are invariably turbulent and may be expressed by $q = Ca\sqrt{2gH}$ in which 'a' is the cross-sectional area of orifice. These distributors are susceptible to clogging. However, still these are used in many cases because they are cheap, easy to insert and not very sensitive to pressure change. In some orifice distributors provisions are there for variable cross-sectional passageways (Fig. 5.7). This provision helps in flushing action at the beginning and end of the watering. At low pressure the comparatively large cross-section passageways opens and allowing higher discharge rate requires in flushing and as the pressure increases a ball or a spring closes the orifice or the slot leading to a decreased rate of discharge. The ball is made of resilient material, which uses to deform at higher pressure and seat lightly on the orifice or slot. Thus, the orifice distributors then adapt the characteristics of compensating distributor.

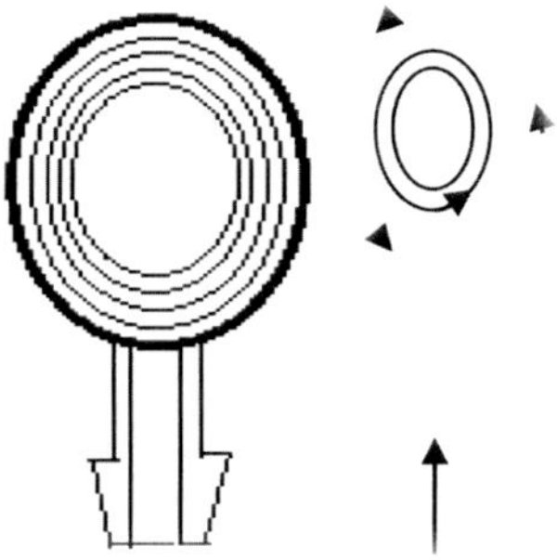

Vortex distributors

It is a simple device in resistance to flow (Fig. 5.8). The water is allowed to enter tangentially in a cylindrical chamber where it is forced to have intensive whirling motion causing much loss of head. The whirled water is passed to a second chamber through an orifice. In this second chamber the water jet is sufficiently broken and drips through the orifice . Vortex distributors are relatively expensive.

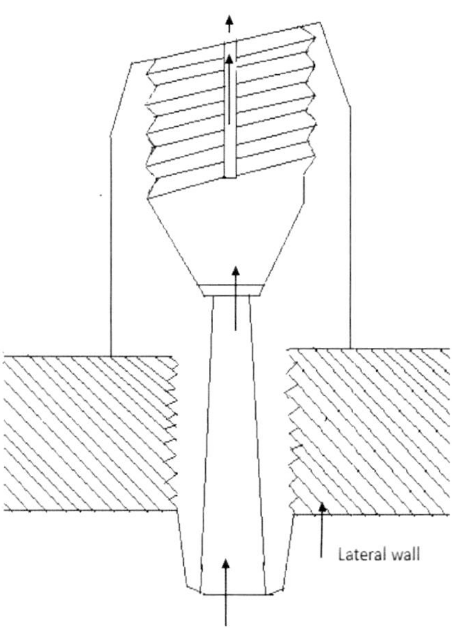

Other type of distributors

Twin-wall distributors

The twin-wall or the dual-chamber distributor is the further development to perforated tubes. This consists of an inner pipe or the supply chamber for conveyance of water and the outer pipe or the emission chamber that emits the water. The inner pipe is having successive openings (orifices) which are comparatively large, the spacing of which depends on the requirement of outflow. The openings in outer pipe are smaller and the ratio of inner to outer opening varies from ¼ to 1/10 depending on the requirement of discharge following the soil characteristics. The pressure in outer pipe is reduced to such extent that the discharge from one opening of inner tube discharges from 3 to 4 outer openings. The pressure in inner pipe is some time as high as 40m, which can be reduced to 0.5m in outer pipe. Thus, these distributors provide the advantage of using high initial pressure followed by low-pressure discharge with comparatively large orifices.

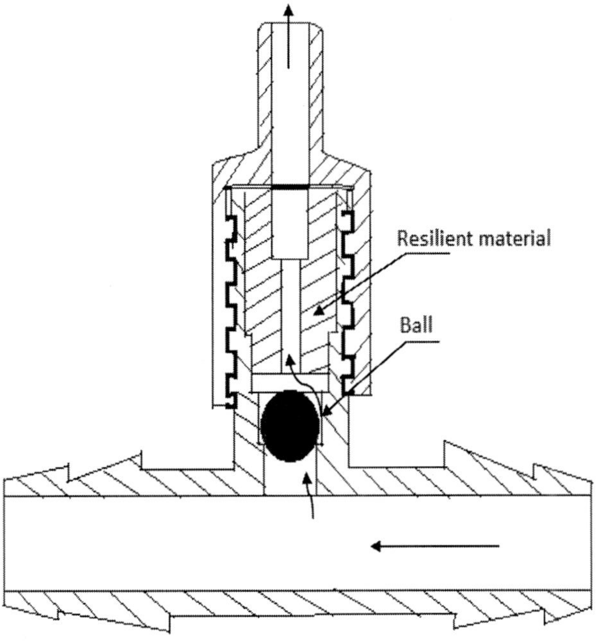

The combined effect of this minimizes the friction loss, permits using these distributors at large variation in elevation and reducing chances of clogging. These distributors are cheap, easy to install and can be used for very low discharge.

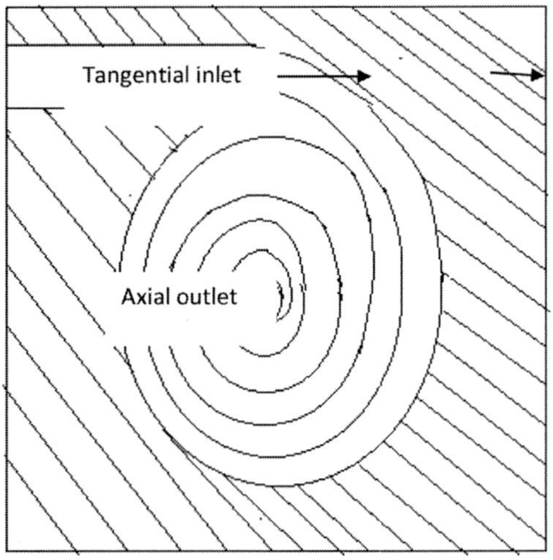

Figure 5.8: Vortex distributor

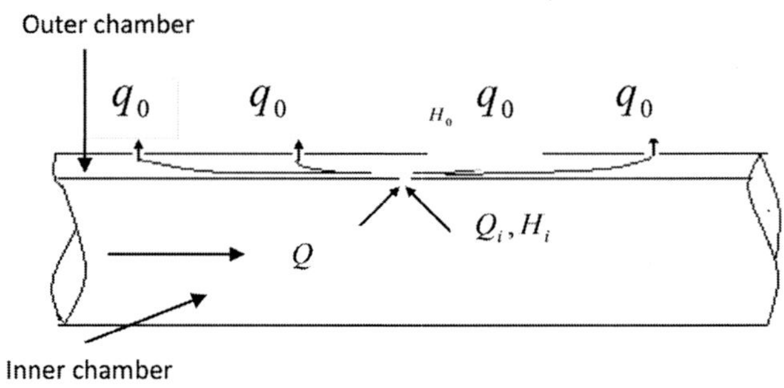

Figure 5.9: Twin-wall system

Referring to Fig. 5.9 the flow equation are given below which may be used for design purpose till a more refined method is suggested (FAO, 1980).

$$H_0 = \frac{H_i}{N^2 (C_0^2 d_0^4 / C_i^2 d_i^4) + 1} \tag{5.2}$$

where, C_0 & C_i = coefficient of discharge through outer and inner pipe respectively

d_0 & d_i = diameter of outer and inner pipe respectively.

N = ratio of the inner orifice to the outer orifice.

Assuming that $C_0 = C_i$ and $d_0 = d_i$

$$H_0 = \frac{H_i}{N^2+1} \tag{5.3}$$

where, H_i = pressure at any point in the inner tube and is equal to the difference of initial pressure (H) and the friction loss (ΔH) from inlet to the point of consideration ($H_i = H - \Delta H$).

H_0 = pressure at any point in the outer tube

$$q_0 = C_0 \frac{\pi}{4} d_0^2 \sqrt{\frac{2gH_i}{N^2+1}} \tag{5.4}$$

The value of C_0 varies with the types of drilled or punched orifices and 0.67 may be a fair value considering the available twin-wall distributors in the market.

Example 5.3 The outer and inner diameter of twin-wall distributors is 2.5mm and 0.25mm respectively. Assuming a pressure of 20m at 30m length of lateral and coefficient of discharges both in outer and inner pipe as 0.67, determine the discharge through the outer pipe.

Solution:

The ratio of outer to inner diameter, $N = \frac{2.5}{0.5} = 5$

The values of C_0 & $C_i = 0.67$

Head at 30m length of lateral, $H_i = 20m$

Head at outer pipe at 30m length,

$$H_0 = \frac{H_i}{N^2(C_0^2 d_0^4 / C_i^2 d_i^4)+1}$$

$$= \frac{20}{5^2((0.67)^2(0.5)^4/(0.67)^2(2.5)^4)+1}$$

$$= \frac{20}{25x\frac{(0.5)^4}{(2.5)^4}+1}$$

$$= \frac{20}{25x1.6^{-0.3}+} = \frac{20}{1.04} = 19.23\text{m}$$

$$q_0 = C_0 \frac{\pi}{4} d_0^2 \sqrt{2gH_i}$$

$$= 0.67x\frac{\pi}{4}x(0.5/1000)^2\sqrt{2x9.81x19.23}$$

$$= 2.55x10^{-6} m^3/s$$

$$= 9.19\text{l/h}$$

Large calibrated orifices with sleeve system

This system consists of polyethylene pipeline of diameter about 25mm having inserted orifice about 2mm diameter covered by a sleeve. The sleeve deflects and breaks the water jet as coming out from the orifice. Since the diameter of the orifice is large, the discharge through it is comparatively large. These distributors are used to avoid the bad effect of flood irrigation. The pipeline and orifice device are laid in small furrow divided into reaches and small ridges separate the reaches. Each reach is served by one orifice. The length of a reach varies 3.5 to 7.0m (Fig. 5.10). The disadvantage of using this device is lying in the fact that uniformity of discharge through the orifices can not be had unless the length of the pipe is short or the diameter of the orifices are adjusted depending on the variation of pressure along the pipe.

Porous wall tubing

Porous wall tubing's are generally buried and therefore belong to the subsurface irrigation system. The porous wall tubing subsurface irrigation may be defined as "application of water below the soil surface at the root zone of the plants through tiny openings provided on the wall of the pipe at a rate that allows the soil to absorb the water at a natural rate"(Suseela *et al.*,2004x). There is little scope of sealing the pores in tubes by the roots when the pipe is buried in soil. These tiny openings are inbuilt pores and are not mechanically made holes. The porous pipes are usually made of recycled rubber and polyethylene. The disadvantage of using porous wall tubing are that the micro pores may become blocked by the contamination of water and in addition to its inherent character of manufacturing difficulties to maintain the uniformity in porosity. Using sand envelope around porous pipe gives better discharge rate and distribution efficiency compared to the porous pipe without sand envelope. The loss of head through the micro pores in pipe wall is large compared to loss of head due to friction loss in flow of water down to lateral. Theoretically the discharge through the porous tubing increases exponentially with the head ($\log q = a + b + H^c$).

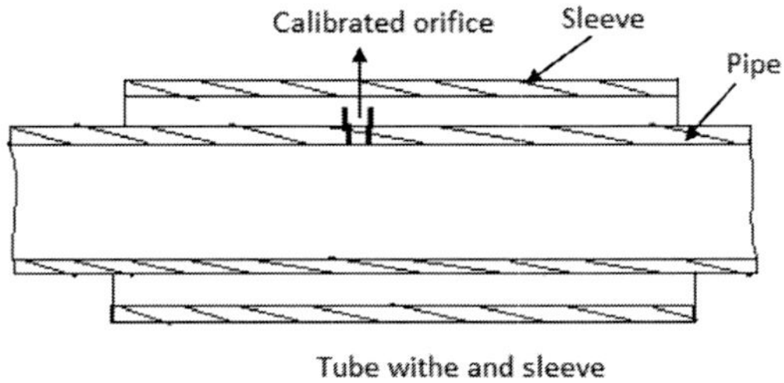

Figure 5.10 Large calibrated orifice with sleeve system

Courtesy: Anonymous (2014)

Figure 5.11: Porous tubing

5.0 Questions & Problems

5.1. Classify the distributors based on the principle of working.

5.2. Describe the method of calculating the method of calculating the length of a micro tube used as distributor.

5.3. Discuss with sketch the working of internal spiral distributor.

5.4. Discuss with necessary sketch the working of twin-wall distributors.

5.5. What are the advantages and disadvantages of large calibrated orifice with sleeve system?

5.6. Determine the length of microtube of diameter 0.6mm at 10, 25 and 75m length for lateral discharge rate of 0.85lph for maintain a pressure gradient of 4%. The initial pressure at the lateral is 12m.

Ans. 1.97, 1.82 & 1.46m

5.7. Determine the length of micro tubes used as distributors of diameter 0.6mm at 30, 40 & 70m length of lateral. The expected rate of flow through the distributors is 0.6l/h. the initial pressure and pressure gradient in lateral is 12m and roughly 4% respectively.

Ans. 2.86, 2.73 & 2.38m

5.8. Develop a chart for lengths of micro tubes of diameters 0.5, 0.6, 0.7, 0.8, 0.9 & 1.0mm at different length ratio of laterals 100, 150 & 200m lengths assuming rate of flow 0.03l/s through the laterals and initial pressure head 15m. Assume any suitable spacing of micro tubes.

5.9. The inner and outer diameter of twin-wall distributors is 0.6mm and 2.5mm respectively. Determine the discharge through the outer pipe if pressure in inner pipe at some length is 15m. Assume 0.67 as the coefficient of discharges. Also determine the discharge through the outer pipe when inner and outer pipe both is of 1.0mm.

Ans. 11.43 & 22.37l/h

5.10. Write True or False of the following

1. The simplest form of orifice distributor is the perforated single chamber tube.
2. The molded distributors can be set and withdraw at any point on the lateral.
3. Porous wall tubing is usually used on ground surface.
4. Drip irrigation is suggested for discharge variation of distributors not more than 5%.
5. Internal spiral molded distributor has the advantage of flushing action or by dismantling the body and cleaning the spiral passage

Ans. 1. True, 2. True, 3. False, 4. True, 5. True

5.11. Select the appropriate answer from the following

1. The distributor in which head loss takes place in principle similar to small diameter tubes may be called as

   a) Twin-wall distributor b) Spiral distributor

   c) Orifice distributor d) Vortex distributor

2. Vortex distributors have

   a) Long path to dissipate energy

   b) Outer pipes to emit water

   c) Cylindrical chamber for whirling motion

   d) Micro tubes coiled around the lateral

3. Drip irrigation system is suggested for emission uniformity not more than

   a) 90% b) 92%

   c) 94% d) 96%

4. The distributor which provides almost same discharge at wide pressure range may be said as

   a) Compensating distributor b) Calibrated orifice

   c) Vortex distributor d) Porous pipe distributor

5. The maximum disadvantages of using micro tube as the distributors are subjected to

   a) Clogging
   b) Displacement inadvertently
   c) Higher cost
   d) Cumbersome to use

6. In calibrated orifice distributor the incoming water jet through the orifice made to drip by

   a) Further reducing the diameter of orifice
   b) Reducing the operating pressure
   c) Breaking the water jet through a baffling action
   d) Temporary stopping the water supply.

7. In twin-wall distributor the ratio of inner to outer openings varies from

   a) ½ to 1/5
   b) ½ to 1/8
   c) ¼ to 1/10
   d) 1/3 to 1/10

Ans. 1. (a), 2. (c), 3. (a), 4. (a), 5. (b), 6. (c), 7. (c)

References

Anonymous (2014).www.Sprinkler-irrigation.co.uk.

Nakayama, F.S. & Bucks, D.A. (1986). Trickle Irrigation for Crop Production: Design, Operation and Maintenance. Elsevier. Amsterdam-Oxford-New York-Tokyo.

Suseela, P. Nagrajan, Rangaswami, M.V., & Palaniswami (2004). Micromising Irrigation Using Porous Pipe. WMRI, Kerala Agricultural University, Vellanikkara & AECRI, Tamil Nadu Agricultural University, Coimbatore.

Thokal, R.T., Mahale, D.M. and Powar, A.G.(2004). Drip Irrigation system: Clogging and Its Prevention. Pointer Publishers, Jaipur.

Vermeiren, I and Jobling, G.A. (1980). Irrigation and Drainage Paper 36, FAO.

6

Fertilization

Introduction

Fertilization is the application of fertilizer for the benefit of the crops. It may be traditional broadcasting or fertigation. In broadcasting method the fertilizer is applied in concentrated granular form. The size of the granules ranges from granular dust to pellet size. These solid fertilizers get dissolved by the rain or irrigated water to leach down the plat root zone. This method results uneven nutrient delivery due to uneven dissolving the fertilizer granules. It may cause burning of plant leaves and root system due to uneven and uncontrolled release of nutrients. Dry fertilizers can volatilize and release gases which may damage or burn the nearby foliage. Availability of fertilizer to the plant root zone largely depends on the temperature and availability of water to dissolve the fertilizers.

In fertigation liquid or water-soluble fertilizer, soil amendments or other products required by the plants is either sprayed on to the plant material or applied to the plant bottom with irrigation water. Foliar application provides high rate of absorption of fertilizer through the leaf/stalk/branch structures of plants. Ground application also provides good consistency in making its way to root zone and plant absorption. Thus, the fertigation provides benefit of increased nutrient absorption; decrease water needs, and reduce the fertilizer and chemical use (Anonyms, 2005).

Drip irrigation provides good opportunity for precision and economic application of fertilizers. The root system of the plants develops extensively in restricted area under drip irrigation. The fertilizers are placed in this high concentrated root zone area along with the water, which favors efficient uptake of nutrients. However, necessary care should be taken to ensure correct amount of fertilizer to the correct place and fertilization do not encourage clogging. Fertigation through drip may not be much efficient unless the flow through the system is steady enough and uniformity of application of water through emitters is within the applicable limit.

Drip irrigation system is suggested for design for discharge variation of distributors not more than 5% from the mean and emission uniformity 94% (Karmeli & Keller, 1975). The chemicals applied in drip system should be completely soluble. There are some chemicals which are coated with clay or wax material to prevent caking in storage. This coating may cause to form sludge to deposit on the bottom of the fertilizer tank. Using the discharge tube from the tank at some height from bottom may allow the safe deposit of the sludge at the bottom and periodical washing prevent it being carried up to the distributors. Injection points to be selected before the filters so that the contaminants get filtered before it reach to the emitters.

Advantage of fertigation

1. Uniformity of application and better absorption: Chemicals are precisely applied with water at the desired location with uniform distribution and greater absorption by the crops.
2. Less expensive: Application cost of chemicals is less expensive due to less labor and energy cost than traditional application.
3. Reduced compaction and crop damage: Compaction and crop damage possibility from fertilizer distribution equipments is avoided.
4. Less operation hazard: Since the operator is not riding on or carrying the system, there is less possibility of contact with chemicals through drip, frequent filling or other exposures.
5. Avoid leaching: Leaching of fertilizer could be avoided by lower doses of fertilizer in split application. Minimize nutrient losses by applying fertilizer around the plants only.
6. Reduce fertilizer requirements: Saving in fertilizer by 25-30% over traditional method of application. Macro and micro nutrients can be applied in one solution with irrigation.
7. Better yield: There is possibility of 25-50% increased yield due to even distribution of fertilizer and water.

Disadvantages of fertilization

1. Higher management practices: Perfect chemical application requires skills and knowledge about the fertilizer, fertilizer equipments and irrigation system.
2. Additional equipments: The injection equipments and safety devices and storage tanks are required.
3. Fertilizer solubility: Readily water soluble fertilizer and liquid fertilizers

are suitable for fertigation. Slowly water soluble fertilizer like super phosphate and calcium ammonium are not suitable for ferigation.

4. Chances of fertilizer precipitation: Some fertilizer like phosphate may precipitate in the pipe line.
5. Corrosion: Equipments need to be made corrosion resistant to its sensitive parts.
6. Crusting soil surface: Chemical application also causes some soil compaction and formation of soil crust.

Fertilizers in Drip Fertigation

Nitrogen

Nitrogen (N) is the major plant nutrient which is applied often in quantity in soil to achieve good production. However, nitrogen availability in soil is limited because in various forms it get leached, utilized, denitrified or fixed in the organic formation of soil (Buck, 1986). There are so many forms of nitrogen available in the market. The major sources of fertilizer-N are anhydrous ammonia, urea, urea sulphate (US-28), urea ammonium nitrate (UAN-32), ammonium sulphate, aqua ammonia, ammonium phosphate, ammonium nitrate and calcium nitrate. The Table 6.1 gives a list of fertilizers along with their solubility and percent of compositions.

The water quality to be taken in to consideration when N is applied in drip irrigation system. Anhydrous ammonia or aqua ammonia when injected to irrigation water it increases the p^H of water which situation is conducive to the precipitation of calcium, magnesium and phosphorous and formation of complex magnesium ammonium phosphates which are insoluble (Nakayama & Bucks, 1986). The situation deteriorates further in the presence of bicarbonate in irrigation water (Rolston *et al.*, 1987). Opposite to the preceding, NH_3NO_3 causes a sharp decrease in soil p^H and soluble aluminum in the wetted zone (Edwards *et al.*, 1982), nitrate salts, such as potassium nitrate or calcium nitrate cause little change to p^H of water.

Table 6.1: Solubility and composition of some commercial fertilizer materials.

Material	**Approximate parts solubility in 100 parts cold water**	**Average percent nutrient composition of materials**			
		N	P	K	Others
Major nutrients:					
Amonium nitrate	118	33.5	-	-	-
Ammonium sulfate	71	21	-	-	-

Calcium nitrate	102	15.5	-	-	21Ca
Diammonium phosphate	43	21	11.5	-	-
Monoammonium phosphate	23	11	10.5	-	-
Orthophosphoric acid	550	-	49	-	-
Potassium chloride	35	-	-	53	-
Potassium nitrate	13	14	-	39	-
Potassium sulfate	12	-	-	45	185
Sodium nitrate	73	16	-	-	-
Superphosphate, single	2	-	4-5	-	20Ca 12S
Superphosphate, Double	4	-	9-10	-	13Ca 10S
Urea	78	45-46	-	-	-
Micronutrients:					
Copper sulfate	22	-	-	-	25Cu
Ferrous sulfate	29	-	-	-	20Fe
Manganese sulfate	105	-	-	-	25Mn
Sodium borate	5	-	-	-	11B
Sodium molybdate	56	-	-	-	40Mo
Zinc sulfate	75	-	-	-	22Zn
Fe-EDDHAa	9	-	-	-	6Fe
Fe-DTPAa	22	-	-	-	lOFe

Ammonium sulphate causes serious clogging and phosphate in it forms complex precipitates, if calcium and magnesium presents in water. Urea is very commonly used for the source of nitrogen. It is very good to soluble in water and reacts little with water to form ions unless enzyme urease is present. However, urease is present in water which contains large amount of algae or other microorganisms. The urease cannot be filtered out and it causes hydrolysis of urea to form ammonium ion.

Phosphates

Phosphate fertilizers are usually not injected in drip irrigation water. Most of the phosphorous fertilizers are susceptible to chemical and physical precipitation problem and clogging to the emitters. The problem of clogging due to phosphorous fertilizer increases further in the presence of calcium and magnesium in irrigation water. However, such problem can be avoided if phosphoric or sulfuric acid can be added to irrigation water. Phosphoric or sulfuric acid helps in keeping the p^H of water low and the salts to remain insoluble (Rauschkolb *et al.*, 1976).

Potassium

Potassium fertigation provides no adverse reaction when they are applied alone in water. The choice of K fertilizer depends on crop needs, crop tolerance, method of application, other elements present in the fertilizer, fertilizer availability and overall the cost. The K content in potassium chloride (KCL) accounts 90% and highest in among all fertilizer formed by K_2SO_4, $K_2SO_4.2MgSO_4$, KNO_3. K_2HPO_4 & KH_2PO_4. Naturally, the selection of a particular fertilizer depends on the availability of materials. Potassium chloride sometime not preferred because its ability to chloride toxicity. The drip system is convenient and efficient for correcting K deficiency by allowing low rates and frequent applications. However, since the drip water is applied to a limited soil volume the salts tends to accumulate in around the periphery of the wetted volume, the highly soluble fertilizers like KCL or many other fertilizers applied in a confined volume of soil and water increases the salinity to a great extent. The extent of development of such salinity needs frequent measurement otherwise significant decrease in growth and yield may occur.

Micro fertigation

Various micronutrient compounds such as sulfate and chelates are generally used for correcting micronutrients deficiencies. Iron and zinc chelates have very good suitability and may have least problems to drippers. Research work is still inadequate in efficiencies of micronutrients applied through drip water. Lindsey & New (1974) reported lower cost in application of zinc chelate in pecan tree through drip system than foliar application. Zinc deficiency in orchards can be corrected successfully by injecting zinc sulfate in to drip water at a less cost than foliar application (Francis, 1977).

Drip Fertigation Systems

Soluble fertilizers may be applied in drip irrigation system in many ways. However, the commonly used system may be grouped as

i. Suction injection system
ii. The pressure differential system
iii. The pump injection system

Suction injection system

It is the simplest method of application of fertilizer in drip irrigation system. As shown in Fig. 6.1 there is a by- pass line of U-shape connected in the main line of the drip system. A venture section is inserted in the middle of the by-pass line. A pipe is connected to the venture to the open supply fertilizer solution tank. When flow occurs through the main and the by-pass line, the venturi section drops pressure by increasing the velocity of flow. This negative pressure draws the fertilizers solution in the tank and inject to the drip system.

Pressure differential system

In pressure differential system, pressure difference is created between the inlet and outlet pipe of the fertilizer tanks by using the pressure reducing valve or the venturi tube in the main line set between the inlet and outlet pipe to the tank (Fig. 6.2). This arrangement causes flow of water in the tank through the inlet pipe and the fertilizer mixed water in the tank to the main water supply line through the outlet pipe of the tank. Thus, the solution constantly is diluted by the new water entering to the tank and displacing the chemicals to the main line. More is the closing of reducing valve, more the water flow from tank to main line. It is necessary to calibrate the tank discharge and the pressure gradient for emptying the fertilizer in the tank in desired time.

Thokal *et al.* (2004) stated the discharges of the tank at different pressure gradient (Table 6.2). In this system, the concentration of the chemicals in the tank changes continuously with the advancement of time. However, there should not be any problem of uniformity in distribution of fertilizer or application of entire fertilizer if sufficient time is allowed for application.

Pressure differential system is suitable in construction and operation, portable, resistant to pressure and discharge variations. However, there is little or no control of chemical injection rate or concentration of chemicals in the irrigation water during the time of operation. The tank should be made of materials protected from possible corrosion and should withstand the required pressure in it. The use of venturi for creating differential pressure in the simplest form permits the use of open fertilizer tank (Fig. 6.1).

The pump injectors

In pump injectors the fertilizer solution in an open tank is injected in the main supply line by using the pump driven by the motor or the water pressure in the main line (Fig. 6.3). In motor driven pump the quantity of injection can be increased or decreased at will. The water pressure driven pump regulates the quantity proportional to the pressure or the quantity of flow in the main line. The pumps are usually rotary, gear, piston or diaphragm type. The fertilizer tank used in the system remains open and usually made of lighter materials. The piston or diaphragm pump injects the chemical solution in to the main line at higher pressure than the pressure in the main. This system is preferred when rated chemical concentration requires to be maintained in the line. Therefore, the system is provided with flow meters, back pressure regulators and flow control valves. The various components have to be well matched to ensure desired flow rate in the drip system.

System hygiene

Fertilizer application through irrigation makes availability of nutrients and possibility of growing more bacteria, algae and slime in the system. Therefore, care should be taken for regular removal it through chlorine or acid. Chlorine should not be used with fertilizer injection since it may tie up the nutrients and cause them to unavailability to the plants. It is necessary to flush of nutrients before completion of irrigation.

Rate of Fertilizer Application

The calculation of rate of fertilizer application is required to avoid the danger of over fertilization. The rate of fertilizer application is site specific and depends mostly on soil, crop, nutrient required, farm area, climate, quantity and quality of water to be applied, fertilizer injector and system design. Application of desired amount of fertilizer in a short time may cause high concentration of nutrients and thereby plant damage. High nutrient concentration leads to toxicity, rapid change in soil p^H and excessive increase in soil salinity of soil and water. In most of the cases the nutrients requirements can be met at a concentration of 100mg/l in drip water. The injection rate of fertilizer through drip may be calculated through the following equation (Nakayama & Buck, 1986).

$$Q_f = \frac{F_r A}{N_c T} \tag{6.1}$$

Where, Q_f = quantity of fertilizer to be applied, m^3/h

F_r = fertilizer rate per application, kg/ha

A = area to be fertilized, ha

N_c = nutrient concentration in the stock solution, kg/m^3

T = period of injection, h

Knowing the projected Q_f the concentration of nutrients in drip water can be determined b

$$C_n = \frac{F_r}{V} \tag{6.2}$$

C_n = concentration of nutrients, kg/m^3

V = volume of water applied, m^3/ha

Having selected the desired concentration of nutrient in the irrigation water, the rate of injection can be determined from the flow rate of the system, density, and percentage of nutrient in the fertilizer by

$$Q_f = \frac{C_n Q}{\rho P} \tag{6.3}$$

Where, Q = rate of flow, m^3/h

ρ = density of fertilizer solution, kg/m^3

P = fertilizer solution, vol/vol

The rate of application is as below when dry fertilizer is used

$$F_d = \frac{CQ}{P} \tag{6.4}$$

Where, F_d = dry fertilizer to be injected, kg/h

C = desired concentration, kg/m^3

Example 6.1 A orchard field is irrigated through drip fertigation. In each irrigation 0.45kgN is applied to each plant with 5mm water. The spacing of plant is 5m × 5m. calculate (a) surface rate of fertilizer application, and (b) concentration of fertilizer in drip water. Assume the wetted area 30%.

Solution:

(a) No. of plants/ha $= \frac{10000m^2}{5m \times 5m} = 400$

Fertilizer applied in overall area/ha = 400 × 0.45 = 180kg

Actual surface rate of fertilizer application $= \frac{180kg}{0.3} = 600kg / Ha$

(b) Concentration of nutrients, $C_n = \frac{F_r}{V}$

$$= \frac{180kg / Ha}{5mm \times 10000m^2 / Ha} = \frac{180 \times 1000000mg}{50000l}$$

$$= 3600mg / l$$

Example 6.2 In a drip fertigation 33% liquid nitrogen of density 1.33kg/l is used with desired concentration of nutrient flow 200mg/l. the flow through the system is 0.5l/s. What is the rate of application of fertilizer?

Solution:

We have, $Q_f = \frac{C_n Q}{\rho P}$

$$= \frac{200mg/l \times 0.5l/s}{1.33kg/l \times 0.33}$$

$$= \frac{2x10^{-4} kg/l \times 0.5l/s}{1.33kg/l \times 0.33}$$

$$= 2.2784l/s$$

$$= 0.82l/h$$

Example 6.3 Dry fertilizer of 46% concentration to be applied with desired concentration of 150mg/l. the flow through the drip system is 0.35l/s. what is the ratio of fertilizer injection?

Solution:

We have, $F_d = \frac{CQ}{P}$

$$= \frac{150mg/l \times 0.35l/s}{0.46}$$

$$= 114.13mg/s$$

$$= 0.41kg/h$$

Example 6.4 A chemical of concentration 20,000 me/l is required to be diluted to 0.5me/l in the main line. Find the required dilution of the chemical in the solution tank if the flow rate ratio of the injector pump and main line is 1:500.

Solution

The required final dilution is 0.5me/l

= 0.5:20.000 = 1:40000

The dilution between injector and supply line = 1:500

Therefore, the chemical in the solution tank to be diluted to 500:40000 = 1:80 i.e., one liter of the concentrated chemical to be diluted in to 79 liters of water to get the secondary stock solution in the tank

6.0 Questions & Problems

6.1. Define fertigation. What are the advantages and disadvantages of fertigation over traditional fertilization?

6.2. Discuss the nitrogen as fertilizer input under drip system.

6.3. Discuss the scope of potassium and phosphate fertilizers in drip fertigation.

6.4. Discuss the scope of micronutrients in fertigation.

6.5. Discuss the pressure differential fertilizer applicator with necessary sketch.

6.6. An orchard field of plant spacing 4.5m × 4.5m is irrigated wih 4mm water. Each plan is provided with 0.25kgN and weed area is 33%. Calculate the concentration of fertilizer in drip water in mg/l.

Ans. 3087.5mg/l

6.7. The concentration of a chemical is 25,000me/l which to be injected in drip water. The flow rate in injector pump and the main line are 5 & 2,000l/h respectively. If the desired solution of the chemical is 1.25me/l in the main line, what is the secondary dilution in the solution tank?

Ans. 1:50

6.8. Select the appropriate answer from the following multiple-choice questions.

1. Drip irrigation system is suggested for emission uniformity not more than

   a) 90% b) 92%

   c) 94% d) 96%

2. Potassium chloride accounts K contents

   a) 50% b) 60%

   c) 70% d) 90%

3. Fertigation saves fertilizer approximately

   a) 20-25% b) 25-30%

   c) 30-35% d) 35-40%

4. Fertigation increases yield approximately

   a) 20-25% b) 25-30%

   c) 30-35% d) 25-50%

5. 200kg fertilizer is applied with 10mm water in 1ha area. The concentration of fertilizer is

   a) 200 mg/l b) 800 mg/l

   c) 1500 mg/l d) 2000 mg/l

Ans. 1. (c) 2. (d) 3. (b) 4. (d) 5. (d)

6.9. Write True or False of the following statements.

1. Drip irrigation is suggested for discharge variation of distributors not more than 5%.
2. Nitrogen availability in soil is abundant.
3. Urea sulphate (US-28) is one of the major sources of nitrogen fertilizer.
4. Phosphate fertilizers are usually not irrigated in drip irrigation water.
5. Clogging due to phosphorous fertilizers increases further in the presence of calcium and magnesium in irrigation water.
6. Potassium fertilizer provides adverse reaction when they are applied alone in water.
7. Pressure differential system provides good uniformity in fertilizer application.
8. Venturi fertilizer applicator should have a closed fertilizer tank.
9. The pump injectors are essentially motor driven.
10. It is essential tocalibrate the tank discharge and the pressure gradient for emptying the fertilizer in the tank in desired time

Ans. 1. True 2. False 3. True 4. True 5. True 6. False 7. True 8. False 9. False 10. True

References

Anonymous (2005). www.ezflofertilizing.com

Nakayama, F.S. & Bucks, D.A.(1986). Trickle Irrigation for Crop Production: Design, Operation and Maintenance. Elsevier. Amsterdam-Oxford-New York-Tokyo.

Francis, L. (1977). Fertilization with drip irrigation: Concepts, practices and problems. Proc. Annual Technical Conf., salt Lake City, Utah, Irrigation association, Silver Spring, Maryland. pp. 80-87.

Karmeli, D. and Keller, J. (1975). Trickle Irrigation Design. Rain Bird Sprinkler Manufacturing Corp., Glendora, California. pp.133.

Rauschkolb, R.S., Rolston, D.E., Miller, R.J., Carlton, A.B., and Burau, R.G. (1976). Phosphorous fertilization with drip irrigation. Soil Sci. Soc. Amer. J. 40: 68-72.

Rolston, D.E., Rauschkolb, R.S., Phene, C.J., miller, R.J., Uriu, K., Carlson, R.M. and Handerson, D.W. (1979). Applying Nutrients and Other to Trickle Irrigated Crops. Univ. of Calf. Bull. 1983, Berkeley, California. pp.14.

Thokal, R.T., Mahale, D.M. and Powar, A.G.(2004). Drip Irrigation system: Clogging and Its Prevention. Pointer Publishers, Jaipur.

7

Low Cost Drip System

Drip system has great promise for efficient utilization of available irrigation water. In our country and elsewhere in the world the scientists are unanimously agreed to adopt drip irrigation method for higher application efficiency and thereby better use of scarce irrigation water. Saksena (1993) reported water losses of only 1-2% in drip irrigation in comparison to 6-9% in sprinkler and 30-35% in surface irrigation method. However, very small area compared to other methods of irrigation so far is irrigated by drip system in our country. Up to 2000 the total area under drip irrigation was 2,59,600 ha in India (Alam & Kumar, 2000). Initial high investment for drip network and pumping system, of course, are discouraging to the most of our farmers. Sharma & Abrol (1993) mentioned that in spite of the very encouraging results and attractive subsidies offered by the Govt. of India, the adoption of the drip irrigation system has not picked up to the expected level. Along with the other reasons they pointed out the high initial cost of the system and lack of research effort on developing cost-effective design suited to local condition are responsible for non-practice of drip irrigation at a considerable scale.

Some research works has been conducted to find out low cost method of drip irrigation. Biswas (1998) developed a manually operated drip irrigation system made of bamboo mains, laterals and sub-laterals. The system and its performance are described below.

Drip Network

The main and laterals of the drip network were made of hollow bamboo pieces each of 3m lengths with 7.5cm and 4.0cm diameter respectively. The sub-lateral or the emitters were the bamboo twigs of 0.25cm inside diameter and 1.05-1.10m length to suit the spacing between the plant rows. The mains and laterals were made for the required length in the field by jointing together the sections one after another (Fig. 7.1). The ends of two sections, which contact each other, were made such that one section easily gets inside the other for about 10-15cm.

Source of water

The source of water was a hand tube well. By manually operating the tube well the water was lifted to the water tank placed on a bamboo platform of 3.0m heights at very close to the tube well (Fig. 7.2).

Figure. 7.1 Network of drip system

Figure. 7.2 Water is lifted to overhead tank

Laying of the drip system

The main of the drip system was made run through the center of the field across the plant lines. All the laterals were originated from the main (Fig. 7.3). The laterals ran through the middle of the plant lines. The sub-laterals made of bamboo twigs were attached with the laterals emitted the water at the plant bottoms. There were two sub-laterals on either side from each point of the lateral (Fig. 7.4).

Application of water

The water tank on the platform was filled up with water by operating the hand tube-well. A flexible polythene pipe was connected from the discharge point of the tank to entry point in the main line. Discharge from the tank was controlled by using the bibcock regulator at the outlet of the tank. The system was allowed to work after filling the tank full.

Figure. 7.3 Laterals originated from a main

Figure. 7.4 Emitters made of bamboo twigs at plant bottom

Data collection

The volume of water in liters was collected from selected sub-laterals in different runs. These data were used to uniformity coefficient and losses of water through the system by using the following formulae (Michael, 1985).

Uniformity coefficient $E_u = 1.0 - \frac{\sum x}{mn}$

Where, m = average value of observations, liters

n = total number of observations

x = numerical deviation of individual observations from the average application rate, liters

(i) Loss of water (%)

$$\frac{\textit{Water discharged from the tank} - \textit{water collected from the sub-lateral}}{\textit{Water discharged from the tank}} \times 100$$

The required average filling and emptying time of the water reservoir, average discharge rate through each sub-lateral or the rate of application of water to each plant, pulse rate of tube well operator, etc., were recorded during the time of operating the drip system (Table 7.1). There were all total 496 numbers of banana plants. Out of these, 320 were irrigated through drip and rest through surface irrigation (Fig. 7.5). The drip and surface irrigations were applied to all the three type of plant varieties to determine the response to irrigations methods. The irrigations were given following 1.0IW/CPE. In drip method the K_p (Pan factor = 0.8), K_c (Crop factor = 1.0) and 50% wetted area were considered to determine the volume of water applied in each irrigation. In surface method the applied water was 5cm in each irrigation.

Table 7.1: Time required for filling and emptying the water reservoir (200 liter capacity)

Height of reservoir, m	Time taken to for emptying the reservoir, min		Discharge/ sub lateral, cm3/min	Pulse rate of tube well operator (No./min)	
	Filling	Emptying		Before	After
3.5	12	48	25.02	60-66	100-110

Figure. 7.5 Developed stage of banana crop under bamboo drip irrigation

Performance of the Drip System

The average of the best performances in respect to uniformity coefficients of emitter discharges was 0.93 (Table 7.2). This was considered good enough in comparison to satisfactory level of 0.85 (Michael, 1985). The loss of water was 9.92% at the best performance of the trials. This was considered higher than the usual losses of 1-25 in a commercial drip system. Assuming the peak pan evaporation rate of 5mm, pan factor = 0.8, crop factor =1.0, percentage of wetting = 50, the water requirement of crop was found 8 l/plant/day. In such situation the operator could have covered an area of 0.2Ha and 500 plants under this bamboo drip system. This area of 0.2Ha was considered as capacity of unit manpower for irrigation in ideal condition. Considering all the costs involvement to place and set the drip network at field the cost of components of drip system were as below:

Table 7.2: Uniformity coefficients and losses of water through the drip network

Side	Trial No.	Uniformity coefficient	Losses of water,%	Average of highest uniformity coefficients of both side	Losses of water in average highest uniformity coefficient
A	1	0.84	33.06	0.93	9.925
	2	0.76	30.76		
	3	0.83	17.54		
	4	0.89	12.95		
	5	0.91	10.32		
	6	0.93	9.53		
B	1	0.74	31.51		
	2	0.81	20.54		
	3	0.84	19.09		
	4	0.89	14.59		
	5	0.92	13.46		
	6	0.92	10.32		

i. Main = Rs.16.50/m

ii. Lateral = Rs.10.00/m

iii. Sub-lateral = Rs.1.70/m

iv. Tube well, water reservoir, reservoir platform, etc. = Rs.2,000.00/0.2Ha

v. Bamboo network (excluding item (iv)) = Rs.3.33/m^2 or Rs.6,660.00/0.2Ha

vi. Drip system = Rs.4.33/m^2 or Rs.8,660.00/0.2Ha

Table 7.3: Performance of the drip system for different type of emitters

Type of emitters	Height of reservoirs	Average discharge, l/h/plant	Distribution efficiency, %
Dripper	2.5	0.42	87.46
Dripper	2.5	1.19	82.13
Dripper	3.5	0.71	84.61
Surgical tube (Knotted)	3.5	0.48	79.61
Surgical tube (Unknotted)	2.5	2.26	69.36
Perforation	2.5	4.79	77.30
Perforation	3.5	5.86	81.55
Perforation (coir winded)	2.5	3.70	71.20
Perforation (coir winded	3.5	4.03	68.50
Perforation (coir winded & subsurface)	2.5	0.64	57.85

During this time of experimentation the market price of PVC pipes used for main, lateral, sub-lateral and dripper were as below.

i. Main (50 mm) = Rs.45.00/m

ii. Lateral (15 mm) = Rs.5.25/m

iii. Sub-lateral (2 mm) = Rs.3.00/m

iv. Dripper = Rs.3.00/piece

Comparing to market rate of the mains, laterals and sub-laterals it appeared that only the cost of lateral was more in bamboo made drip system. The main and sub-lateral were much cheaper. In conventional drip system is conventionally required at the end of each sub-lateral, which could have entirely omitted in drip system under the study. This reduced the cost at the rate of Rs.3.00/plant. The cost of network of the bamboo drip system was found to be less by 26% compared to the cost of commercial drip network (Rs.45, 000/Ha for plant spacing 2mx2m) as reported by INCID (1994). The saving in cost could have been higher than this if it was compared the prevailing rate in the time of study in 1997. The present study was conducted in a place where the growth of weeds of strong root system and activities of termites and other burrowing animals were high. In such a condition longevity of such drip system was expected 3-4 years. It was recommended as the useful drip system to small and marginal farmers for irrigating the row crops in the area of scarce water or soils of high infiltration characteristics.

Figure. 7.6 Main and lateral laying

Figure. 7.7 Laterals, sub laterals and drippers

Figure. 7.8 Water drops from dripper at plant bottom

With the similar set up of a manually operated drip system, Kundu & Mondal (2000) replaced the main and laterals by PVC pipe and tested the performance of five types of water emission systems, viz; dripper, surgical tube, perforation, perforation (coir winded) and perforation (coir winded & subsurface). Highest distribution efficiency (87.46%) was found with the drippers followed by perforation (81.53%) and surgical tube (79.61%). The emitters-perforation (coir winded) and perforation (subsurface and coir winded) recorded in between 57.85-77.30% of distribution efficiency were beyond the acceptable limit of 85% (Table 7.3). Higher distribution efficiency was expected from the properly adjusted knots with the surgical tubes with the attention of frequent removal of iron deposition in it.

Figure. 7.9 Water drops from surgical tube at plant bottom

Table 7.4: Cost of manually operated drip irrigation system using different types of emitters for 0.24Ha

Sl.No.	Types of emitters	Cost of the system (Rs.)
1	Dripper	33083
2	Surgical tube	20359
3	Perforation	27089
4	Perforation (coir winded)	27189
5	Perforation (coir winded and sub surface)	27239

The average time for filling and emptying the water reservoirs at 2.0m and 3.0m heights were found 9 min and 1 hour 15min and 10min and 1 hour 10min respectively. Assuming the peak evaporation rate of the region = 5mm, pan factor = 0.8, crop factor = 1.0, percentage of wetting = 50, the water requirement of crop was estimated= 2 l/m^2. The operator could operate the tube well for 4 hours in different time of the day that enabled him to fill up the reservoirs for 24 times having total volume of water = 4800 liters. This volume of water could cover an area of 0.24ha. Therefore, at an ideal condition the coverage by unit manpower was 0.24Ha. Having the manpower engaged in two shifts each of 8 hours in a day, the area could have covered just double of the previous estimation, i.e., 0.24×2=0.48Ha.

Figure. 7.10 Water is being pumped to reservoir

The cost of the drip system for unit area of 0.24Ha by using drippers was found maximum, (Rs. 33083) and minimum for surgical tube (Rs.20359) (Table 7.4). Thus, either the hydraulic performance or the economy, the surgical tube emitters were the bests among all the types of drippers. The cost of the system could have reduced if less costly accessories such as (instead of GI) gate valves, pipes, elbows etc. were used. The system of drip irrigation studied by Kundu & Mondal (2000) proved to be more cost effective other than its beauty in using the human power by avoiding the non-renewable fossil fuel.

7.0 Questions & Problems

7.1. Discuss the importance of cost-effective drip system in Indian context.

7.2. Discuss the principle of working and construction of a bamboo made low cost drip irrigation system.

7.3. Banana crop is grown in 1ha field with 2m × 2m spacing. The lateral lines pass through the middle of plant rows. Each plant is provided with a sub laterals. What are the length of laterals and sub lateral in this field?

Ans. Lateral = 2500m, Sub lateral = 2500m

7.4. The discharges of 4 emitters were recorded 2.17, 2.23, 1.89 and 2.1l/h. What is the uniformity of coefficient of emitter discharges?

7.5. There were 130 numbers of bamboo twig sub laterals in a field delivering average discharges of 2.2l/h. A overhead water of 180 liters was the source of water. The water tank gets filled and emptied in 10 and 50 minutes respectively. If the drip system operates for 2 hours, what is the percent of water loss?

Ans. 20.55%

7.6. Answer True or False of the following.

1. Bamboo made network of drip system is usually costlier than conventional drip system.
2. Low area drip system can be operated by operating pressure head as low as 2m.
3. Hand tube wells can be used to fill up the overhead water reservoir for drip system.
4. The low diameter pipe like surgical tube can be used as drip distributor.
5. Bamboo made drip network is not subjected to weathering effect.

Ans. 1. False 2. True 3. True 4. True 5. False

References

Anwar Alam & Aswani Kumar (2000). Micro-irrigation system-past, present and future. Proceedings of International Conference on Micro and Sprinkler Irrigation Systems, 8-10 Feb., 2000, Jain Irrigation Hills, Jalgaon, Maharashtra, India. pp.1-17.

Biswas, R.K.(1998). Performance evaluation of a manually operated drip irrigation system made of bamboo mains, laterals and sub-laterals. J. Soil Water Conserv. 42(3&4):162-167.

Kundu, K. & Mondal, P. (2000). Development of A Low Cost Manually Operated Drip Irrigation System An unpublished B.Tech. thesis submitted to Bidhan Chandra Krishi Viswavidyalaya, Faculty of Agril Engg., Nadia, West Bengal.

Saksena, R,S. (1993). Sprinkler and drip irrigation in India- present bottleness and suggested measures for speeder development Proceedings of Workshop on Sprinkler and Drip Irrigation Systems, held during 8-10 December 1993 at Jalgaon, India. pp.26-40.

Samuel, Jose C. and Singh, H.P. (1998). Current trends of micro-irrigation development in Indian horticulture. Proceedings of Workshop on Micro Irrigation and Sprinkler Irrigation System, held at CBIP, New Delhi, during 28-30 April, 1998. pp.1-13.

Sharma, B.R. and I.P.Abrol (1993). Proceedings of Workshop on Sprinkler and Drip Irrigation Systems, held during 8-10 December, 1993 at Jalgaon, India. pp.21-23.

8

Sprinkler Irrigation

Sprinkler irrigation is an advanced method of irrigation in which water is sprayed to air and allowed to fall on the ground similar to rainfall. The spraying of water occurs through nozzle connected to a network of pipes with water under pressure. The rate of application and area of coverage under a sprinkler is regulated by suitable selection of nozzle size and pressure in the system.

History of Sprinkler Irrigation

The sprinkler irrigation has started elsewhere in the world in the early part of twentieth century. Before 1920, sprinkler irrigation was limited to orchard, nurseries and intensive vegetable cultivation. The cost of sprinkler reduced considerably by the development of impact sprinkler and lightweight steel pipe with quick couplers in the 1930s. As a result, sprinkler irrigation began to spread in larger area and to the area of field crops. However, sprinkler irrigation developed mainly after the Second World War with the introduction of lightweight portable aluminum pipe.

Sprinkler irrigation has started in India from mid of 1950s. Due to requirement of high initial investment, the average Indian farmers cannot afford the system and therefore the area under sprinkler irrigation in India was 0.66mha (approx.) out of total irrigated area of 87.80mHa in 1995 (INCID). The sprinkler irrigation was first introduced in India in hilly regions such as Western Ghats in Kerala, Tamil Nadu, Karnataka and in the North Eastern States especially by plantation owners for irrigating tea, coffee and cardamom crops. Since the cost of aluminum pipe is high, to replace it, High Density Polyethylene Pipe (HDPE)/PVC with suitable modification have been introduced in sprinkler system. However, the area under HDPE/PVC piping system is below 20% to the total area under sprinkler irrigation. The remaining area is cultivated under aluminum or steel based sets. The indigenous manufacturing of the system has started in India only 35-40 years ago and there are about fifty numbers of manufacturers till 1996 (INCID). The details of the cultivated area under sprinkler irrigation system in the world and in Indian States are given in Table 8.1 & 8.2 respectively.

Subsidy Scheme to Sprinkler Irrigation in India

The Ministry of Agriculture (GoI) adapted the National Pulse Development Project (NPDP) and Oil Seed and Pulse Development Project (OPDP) during the VII plan period with subsidy to sprinkler irrigation system. In 1992-93 the subsidy was 50%, 75% & 25% of the cost for small & marginal farmers, farmers of SC/ST communities and other category of farmers respectively. This subsidy schemes continued up to VIII plan period with the modification of subsidy at the rate of 50% of the cost for small and marginal farmers only to a maximum of Rs.10,000/- per beneficiary. In 1996-97 the subsidy scheme was further revised as: (a) 90% of the total cost or Rs.25,000 per Ha whichever is less, for small and marginal farmers, SC/ST and women farmers; (b) 70% of the total cost or Rs.25,000 per Ha whichever is less, for other category of farmers. The status of sprinkler irrigation development during VIII plan period is given Table 8.3.

In the year 2006, Ministry of Agriculture (GoI) has initiated the Micro Irrigation (MI) scheme. The key features of the scheme are listed below (Anonymous, 2006).

1. It will be Centrally Sponsored Scheme under which out of the total cost of the MI scheme 40% will be borne by the Central Govt., 10% by the State Govt. and the remaining 50% will be borne by the beneficiary either from his/her resource or soft loan from financial institution.
2. Assistance to farmers will be covering a maximum area of 5ha per beneficiary family.
3. Assistance to drip and sprinkler irrigation will be 75% for the cost of maximum area of 0.5Ha per beneficiary which mill be met entirely by the Central Govt.
4. The Panchayat Raj Institutions (PRIs) will be involved in selecting the beneficiaries.
5. All categories of farmers are covered under the scheme. However, it needs to be ensured that at least 25% of the beneficiaries are of Small and Marginal farmers.
6. The focus will be on the horticultural crops being covered under the National Horticultural Mission. A cluster approach will be adopted.
7. The scheme includes both drip and sprinkler irrigation. However, sprinkler irrigation will be applicable only for those crops where drip irrigation is uneconomical.

8. There will be a strong HRD input for the farmers, field functionaries, seminars/workshops at extensive location to develop skills and improve awareness among farmers about importance of water conservation and management.
9. The Precision Farming Development Centres (PFDC) will provide research and technical support for implementing the scheme.
10. At the national level, National Committee on Plasticulture Application in Horticulture (NCPAH) will be responsible for coordinating the Scheme, while the Executive Committee on Micro Irrigation (ECMI) will approve the Action Plans. At the State level the State Micro Irrigation Committee will coordinate the programme, while at the District level the District Micro Irrigation Committee will oversee the programme.
11. The Scheme will be implemented by an Implementing Agency (IA) appointed by State Govt., which will be District Rural Devlopment Agency (DRDA), or any identified Agency, to whom funds will be released directly on the basis of approved district plan for each year.
12. The DRDA shall prepare Annual Action Plan for the District; get it forwarded by the DMIC and SMIC for approval by the Executive Committee (EC) of NCPAH.
13. Payment will be made through crossed cheque. If the cheque is in the name of the system supplier, the same will be delivered through the farmers/beneficiary.
14. Registration of System Manufacturers will be done by the SMIC for the area of districts.
15. Supply of both good quality systems for drip and sprinkler irrigation having BIS marking, proper after sale service to the satisfaction of the farmers is paramount.

Table 8.1: Country-wise area under sprinkler irrigation

Sl. No.	Country	Year	Area under sprinkler irrigation (ha)	Remarks
1	Afghanistan	1967	1,14,000	
2	Algeria	1994	40,000	
3	Angola	1980	11,445	
4	Australia	1980	10,970	South Australia
5	Austria	1980	46,000	
6	Bahrain	1994	130	

7	Belgium	1980		Sprinkler used for flowers, garden and nursery
8	Benin	1994	4,470	
9	Botswana	1992	892	
10	Brazil	1980		Use started in 1950, mainly for coffee plantation
11	Bulgaria	1980	4,96,000	Around 50% area
12	Burkana Faso	1992	3,900	Around 50% area, humid area
13	Canada	1980	65,000	
14	Congo	1993	111	
15		Cyprus	1980	6.690
16	Czechoslovakia	1980	45,207	
17	Denmark	1980	2,95,000	
18	Ecuador	1980	16,000	
19	Egypt	1993	3,13,000	
20	Germany (Democratic Republic)	1980	4,37,000	
21	Germany (Democratic Republic	1980		Sprinkler for anti-freeze irrigation works
22	Ghana	1994	580	
23	Greece	1980	3,80,000	
24	Guinea	1994	1,594	
25	Hungary	1980	3,33,802	
26	France	1980	6,00,000	
27	India	1995	6,58,000	
28	Iran	1993	47,200	
29	Italy	1980	5,17,000	17% of irrigated area. Upland area irrigated by sprinkler
30	Jordan	1991	5,700	
31	Kenya	1992	21,000	
32	Kuwait	1994	600	
33	Kyrgyzstan	1990	1,41,000	
34	Lebanon	1993	21,000	
35	Libya	1990	4,70,000	
36	Malawi	1992	11,300	
37	Mali	1989	100	

38	Malta	1989	150	
39	Mauritius	1995	14,600	
40	Morocco	1989	1,03,200	
41	Namibia	1992	1,845	
42	Nicaragua	1980	16,460	
43	Norway	1980	69,500	
44	Oman	1993	1,640	
45	Saudi Arabia	1992	10,29,000	
46	Switzerland	1980	To protect crop against frost	
47	Syria	1993	30,000	
48	Taiwan	1980	Orchards and gaarden	
49	Tchad	1988	3,200	
50	Tunisia	1991	55,000	
51	Turkey	1994	2,63,849	
52	Uganda	1980	121	
53	UAE	1993	3,748	
54	USSR	1980	60,00,000	35% of irrigated area. Maldovia, Ukraine, Russian Federation
55	UK	1980	1,17,000	90% of total arae
56	USA	1991	85,72,621	35% irrigated area
57	Yemen	1994	350	
58	Yugoslavia	1980	49,192	35% irrigated area
59	Zimbabwe	1980	1,10,000	70% irrigated area

Source: INCID (1998)

Table 8.2: Area under sprinkler irrigation in different States of India

States	Area under sprinkler irrigation (ha)
Assam	90,000
Andhra Pradesh	17,090
Bihar	160
Gujarat	27,740
Haryana	83,600
Himachal Pradesh	70
Jammu & Kashmir	30
Karnataka	41,900
Kerala	5,800
Madhya Pradesh	1,49,980

Maharashtra	33,120
Orissa	400
Punjab	200
Rajasthan	47,850
Tamil Nadu	32,130
Uttar Pradesh	7,360
West Bengal	1,20,040
Other States + UTs	500

• *Mainly for plantation crops*

Source: INCID (1998)

Table 8.3: Development status of sprinkler irrigation during VIII plan period (1992-97)

States	**Number of sprinkler sets installed**					
	92-93	**93-94**	**94-95**	**95-96**	**96-97**	**Total**
Andhra Pradesh	809	1159	3791	7652	5740	19151
Gujarat	1020	1597	2658	5514	2105	12894
Haryana	652	NA	712	1964	760	4088
Karnataka	1524	1202	927	1541	1310	6504
Madhya Pradesh	146	307	1068	8576	6530	16627
Maharashtra	2131	2577	2996	6081	4755	18540
Rajasthan	907	2269	6360	13187	7030	29753
Tamil Nadu	1988	2221	3477	3125	1264	12075
Uttar Pradessh	343	809	3583	3278	3600	11613
Other States	61	143	43	1344	1659	3250
Total	9581	12284	25615	52262	34753	134495

Source: INCID (1998)

Advantage and Limitations of Sprinkler Irrigation

The purpose of irrigation is to provide the required amount of water for maintain the desired soil moisture regime in the root zone of the crop. The irrigation has to be done at a reasonable cost, power and labor. The advantages of sprinkler irrigation over the surface irrigation may be summarized as below:

Advantages

1. Suitable to almost all soils of infiltration rate less than 4cm/h.
2. Suitable to almost all crops.
3. Suitable to uneven land. Land leveling is not required.
4. Fertilizer, herbicides & fungicides can be applied in irrigation water economically.

5. Can be used against protection in winter frost and for cooling of crops in summer.
6. Supply channels and bunds are not required.
7. Saving in water and labor.
8. Permits movement of farm machinery.
9. Healthy growth of crops and higher yields.
10. Less infestation of pests and diseases.

Limitations

1. Wind distorts sprinkler pattern and cause uneven distribution.
2. Ripening soft fruit must be protected from impact of spray.
3. Higher initial cost excepting where high cost is involved in land leveling.
4. High power requirement as it is operated at 5m to 100m head of water.
5. Not suitable for fine textured soil of slow infiltration rate.
6. Movement of portable pipes in some soils after irrigation may pose a problem.

Other Use of Sprinkler System

Sprinkler system has numerous other uses. The following are some of them.

1. Cooling cold storage, livestock and poultry environments.
2. Farm fire protection.
3. Water distribution for compaction of earth fills.
4. Setting of dusts.
5. Log curing.

Scope of Sprinkler Irrigation in India

There is ample scope of bringing the crops like millets, pulses, gram, wheat, sugarcane, groundnut, cotton, vegetables and fruit flowers, spices and condiments under sprinkler irrigation. These crops occupy the vast area (Table 8.4). Thus, there is great possibility to increase the area in phase wise with required financial and technical support from the concerned agencies.

The groundwater supplies is depleting day by day while demand for water is ever increasing. The water crisis is a major issue particularly during the summer season in many parts of the country. Therefore, reduction of groundwater use without compromising the production is a standing issue, which will remain persistent all through with us. More and more use of sprinkler irrigation may provide good scope to attend the problem if necessary political will; policy support and organizational efforts, all together come forward. Sprinkler irrigation is also having a positive impact to environment.

Type of Sprinkler System and Components

The sprinkler system may be divided in to two major types depending on the arrangement of spraying irrigation water.

Table 8.4: Potential areas for sprinkler irrigation in India

Sl. No.	Crop	Area (mha)
1	Cereals and Millets (excluding rice)	27.6
2	Pulses	4.2
3	Oil seeds	11.1
4	Cotton	2.6
5	Condiments & spices	1.2•
6	Fruits & vegetables	2.5•
7	Sugarcane	3.3

• Up to 91-91. For the other crops, cropped area is up to 94-95 and coverage under irrigation is up to 92-93.

Source: INCID (1998)

Figure 8.1 Sprinklers are in operation in field crops

Courtesy: INCID, 1988

i. Rotating head
ii. Perforated pipe

Rotating head system: This method consists of the nozzles set on the riser pipes and the riser pipes on the laterals at regular intervals. The laterals are usually placed on the ground surface. The height of the riser pipes is such that the nozzles on it can effectively spray the water above the crop height (Fig. 8.1). Nozzles are usually used in pair in a sprinkler set opposite to each other. On rotating the nozzle by 90° it covers the area in circular form with some overlapping by the nozzles surrounding it.

Perforated pipe system: In this system the lateral pipes are made perforated in a definite pattern at a regular interval to allow the water come out and distribute to a fairly uniform rate. The pressure in the laterals is maintained much low (<1.5kg/cm^2). Thus, the sprinklers can be operated by connecting the system to an overhead water tank. The perforations are made from two opposite sides of the laterals. The sprays from both the laterals cover the strip of land of width 5 to 15m depending on the pressure in lateral. The rate of application in perforated pipe system is usually high. Therefore, it is suitable for soils of infiltration rate moderate to high. This is mostly used in irrigating the lawns and gardens or vegetables where plant height is less (<60cm).

Classification Based on Portability

The sprinkler system are classified in to the following based on the portability.

i. **Portable system:** the system contains the portable mains, sub mains, laterals and pumping unit. The system is carried from one field to another to irrigate the fields. Off course, each field should have the water source. The portable system is usually used for two or three supplemental irrigation. The system may be carried either manually or mechanically. Manual or the *hands move system* requires much labor to handle it. It is also tedious but requires less initial investment. The system using mechanical power to move the system called as *wheel move system*. The laterals are mounted on the wheels and the set as a whole moves. The initial cost of this is much higher than the hand move system.

ii. **Semi-portable:** A semi-portable system has the portable laterals, mains and sub mains but the water source and pumping unit is fixed. In this system, a few fields may be irrigated by shifting the mains and laterals but the extended main to be in connection to the permanently located pump.

iii. **Solid-set system:** In solid-set system the laterals remain fixed in the field for a crop season. In fact, this is a permanent system for a season. Solid-set system is used where frequent irrigation is required at short interval.

iv. **Semi-permanent system:** In semi-permanent system all are fixed excepting the laterals. The laterals are carried to the fields to connect it to the main or the sub main already available at there.

v. **Permanent system:** A permanent system should have fixed main, sub mains, laterals, water source, and pumping unit. The main, sub mains and laterals are usually buried in soils and the nozzles are set on the risers. The initial cost of this system is high but suited to automation. It is suitable for irrigating the orchards.

Component of Sprinkler Irrigation System

The typical sprinkler irrigation system consists of the (i) Pumping unit, (ii) Pipe net work-mains, sub mains and laterals, (iii) Sprinkler head, and (iv) Accessories such as couplers, valves, plugs, risers, and fittings.

Pump set

The pump is required to lift the water from the source to the supply line at some desired pressure so that the sprinklers at the remote point also receives sufficient pressure to sprinkling water and provide good uniformity to irrigation. The pumping plant of the sprinkler system usually consists of centrifugal or turbine pump, a driving unit, a suction line and a foot valve for the centrifugal pump.

The centrifugal pump is more popular and used in the situation when the source of water is an open well or shallow depth water source less than 8m. The turbine pump is used in deep tube wells. The centrifugal pump is usually set on a trolley and portable but the turbine pump is fixed in a suitable location. Internal combustion engine is preferred to centrifugal pump because of its portable characteristics and electric motor to turbine pump because of its fixed installation. However, in some field level situation users are become compelled to use internal combustion engine to turbine pump.

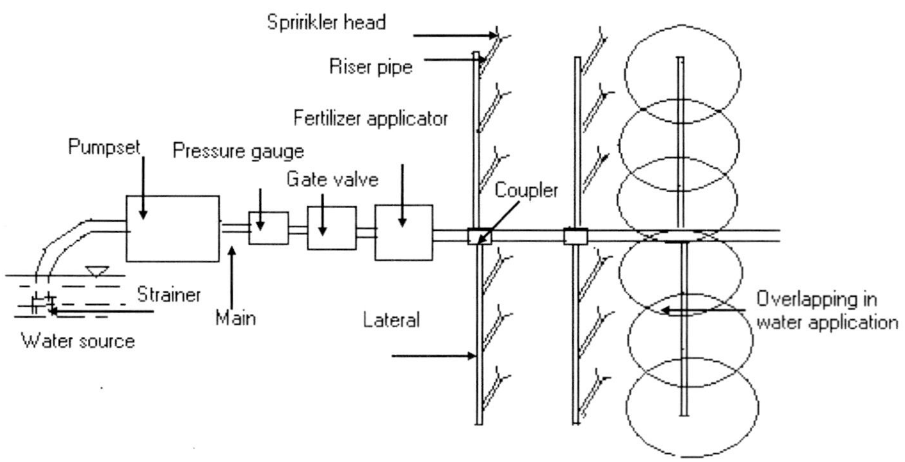

Figure 8.2: Sprinkler irrigation system

Pipe network

The pipe network consists of main, sub-main and laterals. The main line carries the water from the pumping plant to different part of the field through sub-mains and laterals. Sub mains are provided to take water from the main and the laterals are from the sub mains or the mains (Fig. 8.2).

Mains lines: The main lines may be used as permanent, semi permanent or portable. However, the laterals are almost portable type. Permanent main is used where the field area is fixed and the irrigation is required throughout the season. The portable mains are used when the sprinkler system has to irrigate a few fields. Steel, plastic, aluminum, asbestos, cement concrete, wrapped aluminum pipes may be used as mains. The steel pipes are mostly used in permanent lines. The asbestos or the concrete pipes are not suitable for high-pressure system. The main lines are usually buried in to 45 to 60cm below the ground level so that it may be out of the impact of farming operations. The lightweight aluminum pipes or plastics (HDPC) with quick couplers are suitable for most portable main lines. The mains and sub main pipes are of same type excepting the diameter of sub mains is less since the discharge of mains getting divided in to number of sub mains.

(a) Sprinkler nozzle (Hallmark HT-46)

(b) Mini sprinkler nozzle

Figure 8.3: Sprinkler nozzles, Courtesy: Hallmark, 2004

Laterals: The laterals are usually portable. However, it may be buried permanent for orchard, lawns or other special sites where cultivation practices do not create hindrance due to permanent placement of laterals. The laterals are of 5, 6 or 12m and each length are equipped with quick coupling devices which enable the farmers to shift it quickly during farming operation.

Sprinkler head

The sprinkler may be classified as rotating and fixed type (Fig. 8.1). The performance of a sprinkler nozzle (Fig. 8.3(a) under ideal conditions of temperature, wind velocity and humidity along with the specification are stated in Table 6.5 (Hallmark, 2004).Sometime, perforated lateral pipes are also used as sprinklers. The sprinklers are adapted to wide application rate and spacing. The operating pressures are usually ranges between 1.5 to 4kg/

cm^2. The most of the sprinklers used in agriculture are slow rotation type. It may cover small to large area with minimum overlapping and application rate inconformity to infiltration rate of the soil provides better use of sprinkler water. The sprinkler may be single and double nozzle. The single nozzle sprinklers are used for low application rate. Double nozzle sprinklers provide higher application rate. One of the nozzles of double nozzle sprinkler applies water to the considerable distance and the other nozzle cover the area near the sprinkler. Thus, a good uniformity of application is achieved. The revolving head sprinklers may be classified based on their pressure range and position of use in relation to the crops (Table 8.6). It is important to ensure the required pressure on the sprinkler head. Too high or too low a pressure will cause much poor distribution efficiency of water.

Table 8.5: Performance of a water sprinkler (HT-46)

Nozzle size (mm)	Pressure (kg/cm2)	Diameter (m)	Discharge (lpm)
4.36 × 2.38	2.00	26.80	22.00
	3.00	29.85	29.00
	4.00	32.30	32.00
	5.00	33.50	35.00
4.76 × 3.17	2.00	28.00	28.00
	3.00	31.15	36.00
	4.00	33.53	42.00
	5.00	34.15	46.00
5.15 × 3.17	2.00	28.70	32.00
	3.00	32.00	40.00
	4.00	34.14	47.00
	5.00	34.88	51.00
5.55 × 3.17	2.00	29.90	37.00
	3.00	33.20	46.00
	4.00	35.45	54.00
	5.00	36.86	58.00

Courtesy: Hallmark, 2004

Specification

- 3/4 inches BSP/NPT male/female threads
- Body, arm, bearing & nozzles made of derlin for durability
- Non clog stream straightener in range nozzles ensures excellent performance in wind condition
- Special bayonet nozzles
- Stainless steel spring and pivot pin
- Trajectory angle 23°

Table 8.6: Classification of rotating head sprinklers, their characteristics and adaptability

Type of sprinkler	Gravity-fed under tree sprinkler system	Normal under-tree sprinkler system	Permanent overhead system	Small overhead system	Low pressure system	Intermediate pressure system	High pressure systems
Pressure range	0.7 to 1 kg/cm^2	1 to 2.5 kg/cm^2	3.5 to 4.5 kg/cm^2	2.5 to 4 kg/cm^2	1.5 to 2.5 kg/cm^2	2.5 to 5 kg/cm^2	5 to 10 kg/cm^2
Sprinkler discharge	0.06-0.251/s	0.06-0.251/s	0.2-0.61/s	0.6-0.201/s	0.3-11/s	2-101/s	10-501/s
Diameter of nozzles	1-6mm	1.5-6mm	3 -6mm	6- 10mm	3 -6mm	10-20mm	20-40mm
Diameter of coverage	10-14m	6-23m	30-45m	25-35m	20-3 5m	40-80m	80-140m
Range of sprinkler spacings (square)	-	-	18-30m	9-24m	9- 18m	24-54m	54- 100m
Recommended speed of sprinkler rotation	-	0.5-1 rpm	1 rpm	0.67-1 rpm	0.5-1 rpm	0.7 rpm	0.5 rpm
Adaptability	Usually uses single nozzle sprinkler heads, used as under tree systems in uplands, has low uniformity of coverage.	Usually used in closed space orchards with full low hanging branches, single nozzle slow rotation sprinklers often used.	Used for orchards. Triangular spacing necessary for low application rates (1.5-3mm/h).	Commonly used for low rate of application (3.5 -6mm/h) and to help reduce the effects of wind. High risers are used for orchards and lower risers for field crops.	Two-nozzle sprinklers can be used with lower pressure than single nozzle sprinklers. More overlap is required. Rate of application tends to be high.	Usually single nozzle sprinklers, rates of application range from 6-12mm/h, suitable from supplemental irrigation, unsuitable under windy conditions.	Used in dense high plantings. Unsuitable where there is appreciable wind.

Source: Michael (1978)

Most of the rotating sprinkler heads operate following the principle of actuating a small hammer by the force of water striking against a vane connecting to it. The giant sprinklers that operate under high pressure rotate on water activated gear drive. Fixed type sprinklers are used to irrigate the small lawns and gardens. The variation of the fixed type sprinkler is the 'pop-up' sprinkler. In this, the sprinkler used is housed in a casing, which is provided with a cover at the top. The sprinkler unit pops up out of the casing when put to use and sinks down in to the casing when not in use. The perforated pipes require less pressure than the rotating sprinklers. However, they release more water than the rotating sprinklers. Therefore, its use is restricted to soils of high infiltration rate. The system is not suitable where the wind distorts the water jet. Its use is also limited to nearly flat land since the rate of discharge through the perforations varies with the pressure variation causes due to difference in elevation. The perforations also get frequently clogged and require periodical flushing and screen at the intake of pump to restrict foreign particles in the incoming water.

Accessories

Pressure gauge: The pressure gauge is used to measure the pressure in main or other points of the system (Fig. 8.4). These are usually bourdon gauge type and fixed in the system. The portable pressure gauge is used to measure the pressure in sprinkler head (Fig. 8.5). The portable gauge-pack with a pitot tube enables the operator to read the sprinkler pressure in the nozzle under use.

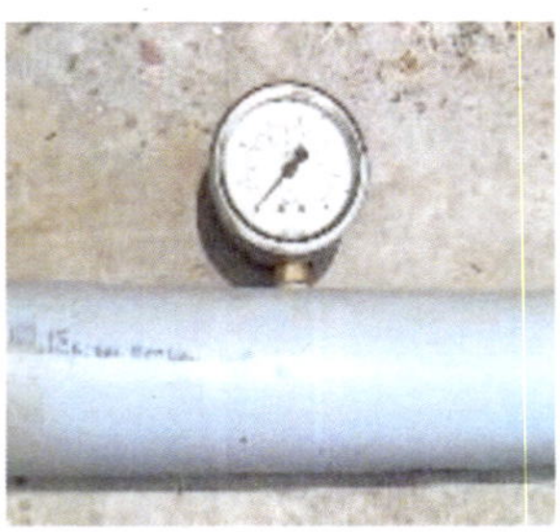

Figure 8.5: Portable pressure gauge

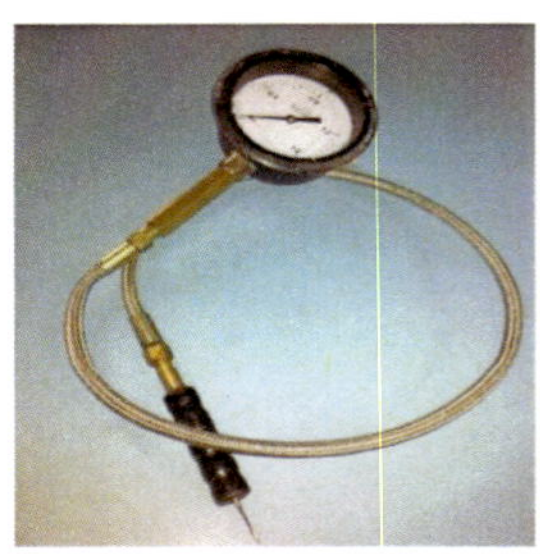

Figure 8.4: Fixed type pressure gauge

Couplers: The couplers provide connection between two pipes and between pipes and fittings. The couplers should have the characteristics of flexible connection, no leak of water through the joint under pressure and automatically drain out at no pressure, simple and easy to couple and uncouple, and overall be light, non corrosive and durable.

Water meter: The water meter is used to measure the water delivered to the main, sub main or to any lateral (Fig. 8.6). The water can records the quantity of water supplied at any instant of time. Thus, it provides the supply of any predetermined amount in a given time to the field. In India the irrigation water is not usually measured as because the users use to pay on area basis. However, this is necessary towards the economical and efficient management of costly irrigation water.

Fertilizer applicator: It is used to apply the fertilizers, which are soluble to water through the sprinkler at a desired rate. The fertilizer applicator is connected usually to the main as a component of control head. The flow through the fertilizer applicator is induced by creating the slight difference in pressure at the water inlet and outlet of it. The difference of pressure causes to suck fertilizer with water in sprinkler system (Fig. 8.7).

Pressure regulators: The sprinkler operation system should be such that all the sprinklers should operate within the pressure variation of acceptable limit. To ensure such operation pressure regulators may be used particularly where excessive variation of pressure occurs due to sloping land and excessive pressure-head losses. This is also important when the pressure head at the water source fluctuate constantly.

Connectors: Flanges, couplings and nipples are used for suitably connect the pump and suction pipes.

Figure 8.6: Water meter

Figure 8.7: Fertilizer applicator
Courtesy: INCID (1998)

The following are the illustration and description of some of the fittings used in sprinkler system by the manufacturer *Rungta Irrigation Ltd., Delhi.*

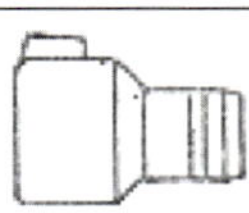	***Main line coupler:*** *To join two pipes together with the help of a hook and clamp assembly.*
	Sprinkler coupler: *To join two pipes together. Has a vertical outlet of ¾" or 1" BSP and a flat base with two holes for holding a batten. The vertical outlet is provided for fixing sprinkler riser pipes.*

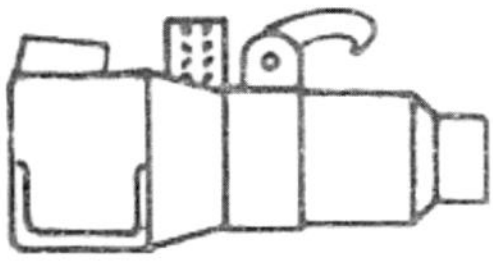	***Sprinkler Adapter:*** *This is similar to Sprinkler Coupler having a vertical outlet for 1" or ¾" BSP. It is an independent fitting and can be used anywhere in the pipeline.*
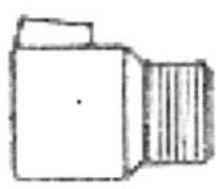	***Screwed Coupler (Female):*** *One end has an external BSP thread for joining G.I. Socket and the other end is a female coupler for joining aluminium pipe.*
	Screwed Coupler (Male): *Provided with BSP thread (external) on one side and male coupler on the other end. Used for connecting aluminum pipe to G.I. pipe/Hose pipe.*
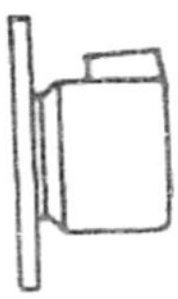	***Flanged Coupler (Female):*** *Provided with a standard flange on one side for joining the pump flange. The other end is female coupler for joining aluminium pipe.*
	Flanged coupler (Male): *Provided with a standard flange on one side and male coupler on the other side for joining aluminium pipe.*
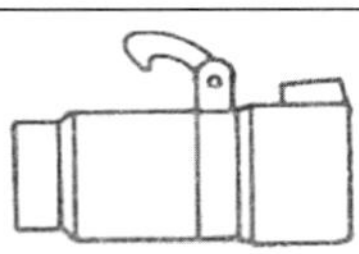	***Reducer Coupler:*** *Fixed to female end of a coupler to reduce the size to next smaller size of pipe.*
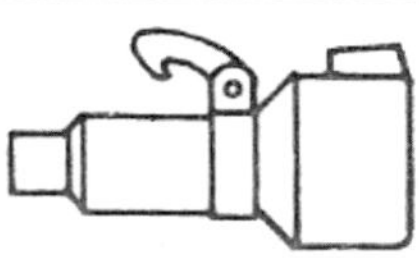	***Increaser Coupler:*** *Fixed to female end to increase the size to next higher size of pipe.*
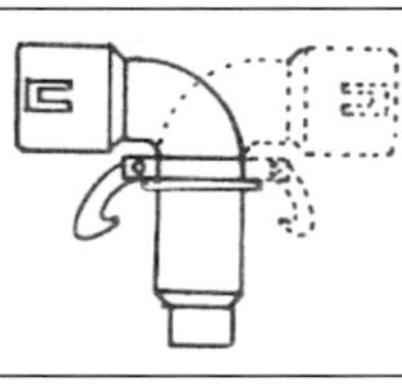	**Bend-90^0:** *This is used for taking a turn in the pipeline at 90^0 angles and has a female coupler and is reversible so as to be left or right hand side.*
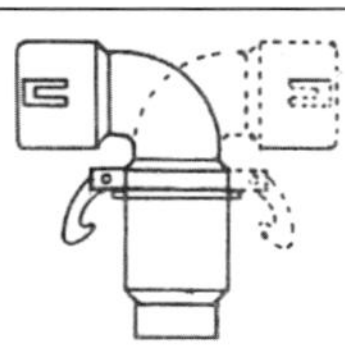	***Reducer Bend-90^0:*** *This is used for joining bigger size pipe to next smaller size pipe at 90^0 angle. This has bigger size male end and smaller size female end. It is also reversible so as to be left or right hand side.*

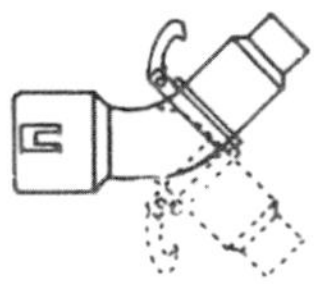	***Bend –45".*** *This is used for taking a turn in pipe line at 45⁰ angle and has a female coupler and is reversible so as to be at left or right hand side.*
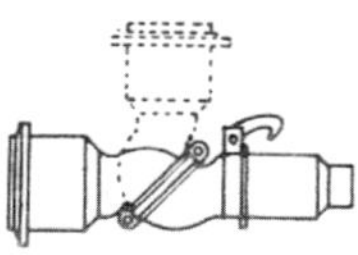	***Adjustable Bend:*** Has a male end on one side to fit inside the female end of a coupler and with a female end on the others side which is adjustable by hand to any angle between 0^0 to 90^0.
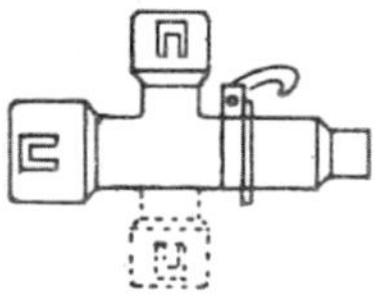	***Side Tee:*** *This is used for joining laterals with the main line on left or right hand side.*
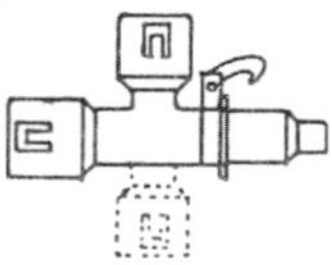	***Reducer Tee:*** *This is used for joining laterals of next smaller size with the main line on left or right hand side.*
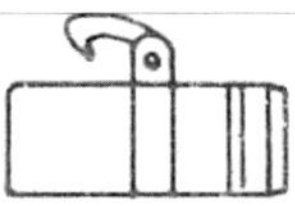	***Male End Coupler:*** *These male end couplers are used for high-pressure heavy-duty installations. These are fitted on the pipe end in place of the clamp and hook normally supplied with couplers. Male end couplers are supplied against specific orders only.*
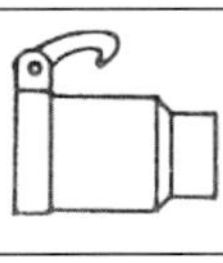	***End Plug or End Stop:*** *To close the female end of the last pipe of mainline or lateral line.*
	***Riser Outlet Tee:*** *Connects into pipelines, one end has male coupler and the other end has female coupler, provided with vertical riser outlet of 75mm (normal dia) to which riser outlet connector coupler can be fitted.*
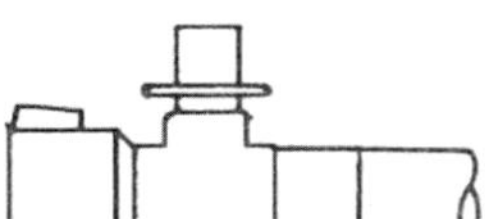	***Main Line Valve Coupler-Heavy Duty:*** *Joins two pipes together, has 75mm or 100mm outlets and heavy-duty adjustable valve. Lateral line is connected to outlet of valve opener. All steel parts are galvanized to protect against corrosion. Valve opener is used to operate the valve.*
	***Insert Valve Coupler-Heavy Duty:*** *This is identical to main line valve coupler but instead of permanently fitted to the aluminium pipe, is provided with a male end coupler to facilitate independent use anywhere in the pipeline like any other fitting.*

	***Riser Outlet Connector:*** *This is used with Riser Outlet Tee for connecting a riser pipe of 11/4" dia.*
	***Valve Opener-Heavy Duty****: Has an inlet part to fit over the outlet in the main line valve coupler or hydrant screwed or inserts valve coupler, which is held on by two hooks that automatically engage with the outlet and can be disengaged by hand. The valve opener will rotate on the outlet through* 90^0*. A hand wheel is attached to the spindle so that by turning the wheel one can open and close the valve and adjust it to any position in between.*
	***Hydrant Screwed-Heavy Duty****: Has an outlet valve assembly like main line valve coupler and at the base, has an inlet connection of BSP internal thread or flange. Valve opener is used to operate hydrant.*
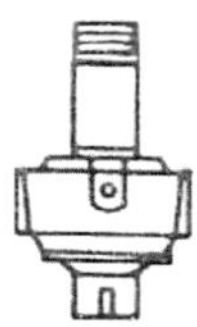	***Ball Joint Riser Coupler:*** *It has a bottom cup, top cup and a hollow ball. The ball outlet can be adjusted in various directions at an angle to the inlet. Inlet has BSP INTERNAL THREAD CONNECTION for fixing a G.I. nipple and the outlet has internal thread connection.*
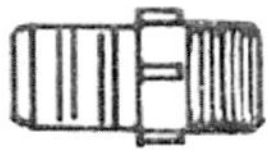	***Screwed Spigot Adapter****: Provided with BSP thread (on external) and is meant for fitting to aluminium pipes permanently.*
	***Sleeve Coupler****: This used for low pressure application. It makes a leak proof joint between the pipes and is generally used for suction line of the pump set.*
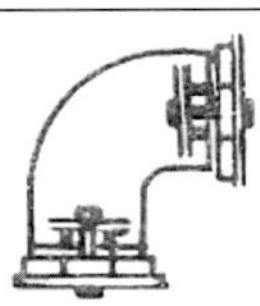	***Sleeve Bend*** 90^0: *This bend provides leak proof joint between pipes for low pressure application and is generally used in suction line of the pump set.*
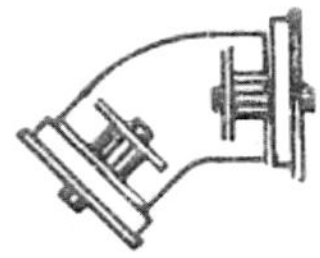	***Sleeve Bend*** 45^0: *This is similar to sleeve bend* 90^0*, but has* 45^0 *instead.*
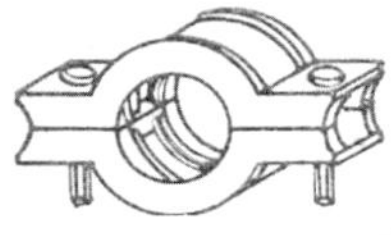	***Groove Clamp Coupler****: Compromises of two half circular clamps which when bolted together completely engage in circumferential groove being made on each end of two pipes joined.*

	***Groove Clamp Screwed Coupler:*** *Has external BSP threads on one end for joining GI socket and the other end has a groove and is used as starter coupler for grooved pipeline*
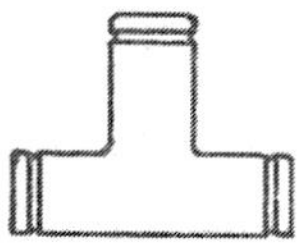	***Groove Clamp Tee:*** *this Tee has all the three ends grooved to fit clamp couplers.*
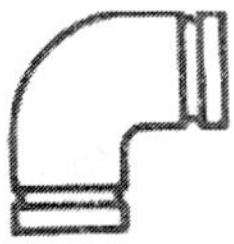	***Groove Clamp 45^0 Bend:*** *This bend having both ends grooved to fit clamp coupler.*
	***Groove Clamp End Stop:*** *To close the ends of groove clamp pipes.*
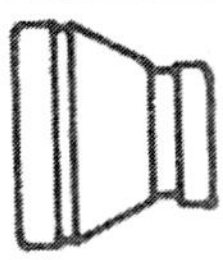	***Groove Clamp Reducer/Increaser:*** *This fitting can be used for reducing or increasing and is used with groove pipes. Both the ends are grooved so that clamp couplers can be fitted.*
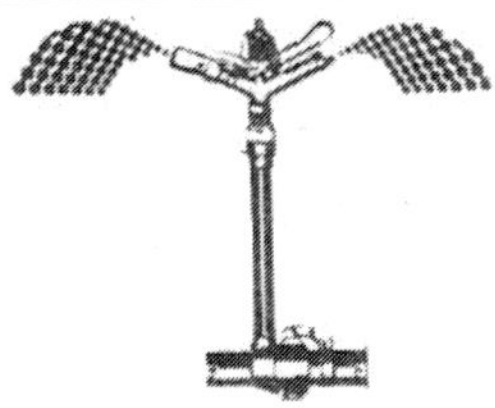	***Sprinklers:*** *Available in various models discharge ranging from 0.79gpm to 250gpm and diameter of spray ranging from 7 ft to 307 ft, having excellent uniformity of application.*
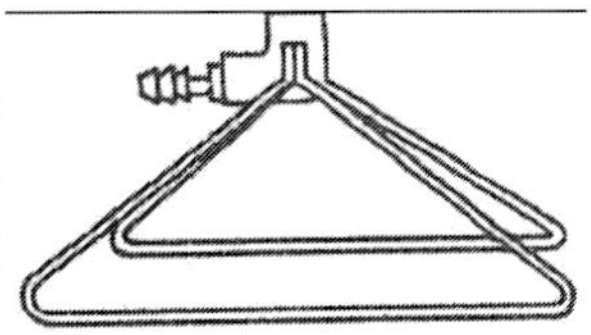	***Garden Stand:*** *Manufactured from MS rod, duly painted for protection against corrosion provided with GI elbow and nipple for connecting rubber hose pipe for irrigating gardens or lawns.*
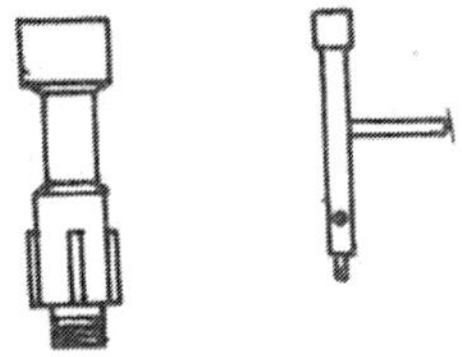	***Quick Riser Coupler:*** *It has an automatic on/off valve and is used with an adopter to operate. This aluminum body part is fitted to the adopter.*

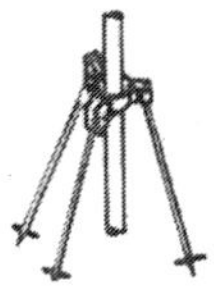	***Riser Pipe with Tripod Stand:*** *It has a GI riser pipe and an adjustable tripod stand to support. One end of the riser pipe is connected to the sprinkler coupler through BJRC/BVC/QRC and a sprinkler is fitted on the top end of the riser pipe.*
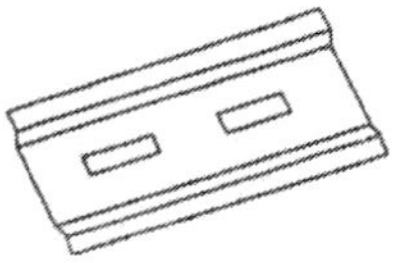	***Batten:*** *Manufactured from GI sheet for fixing to the flat base of sprinkler coupler. This is used when the height of the riser pipe is less than 36 for stability.*

Principle of operation of a rotating impact sprinkler

The rotating impact sprinkler has almost versatile use. It is single or double nozzle and may be used in semi-portable, portable and solid-set system and even to some movable machine. It can operate under wide variation of operating pressure, discharges, spacing and application rates to suit the requirements of different crops and soils.

Principles of sprinkler operation

The water jet ejected from the nozzle of the sprinkler break down in to different diameter droplets over an area around each sprinkler. The principle of sprinkler application of water is to apply uniformly the calculated depth of water at a predetermined rate. The distribution pattern is the representation of water depths in different points within the area of coverage. This can be measured by collecting the water in the cans spaced evenly in the area of application. The distribution pattern is symmetrical at no or light wind condition but get disturbed under high wind. The characteristic distribution pattern of the sprinklers is generally provided by the manufacturers under no wind condition. The spray distribution characteristic of a sprinkler varies much with nozzle size and operating pressure. At lower pressures, the droplets are larger and fall away from the sprinkler. At higher pressures, the droplets are much finer and fall near the sprinkler. Finer particles are susceptible to excessive wind-drift and evaporation. Thus, in case of much variation of pressure, either low or high, there is great scope of poor application uniformity though the sprinklers are spaced properly. This necessitates the appropriate design to ensure the suitable pressure in every point in the field.

Distribution pattern: The typical distribution pattern of single and double nozzle sprinklers are trapezoidal and triangular respectively as shown in Fig. 8.8.

The probable distribution pattern in respect of depth of water and distance from the sprinkler under favourable and windy conditions are shown in Fig. 8.9. Since the water application rate of sprinkler is always less or equal to intake rate of soil the applied water is taken to be absorbed at the point of application.

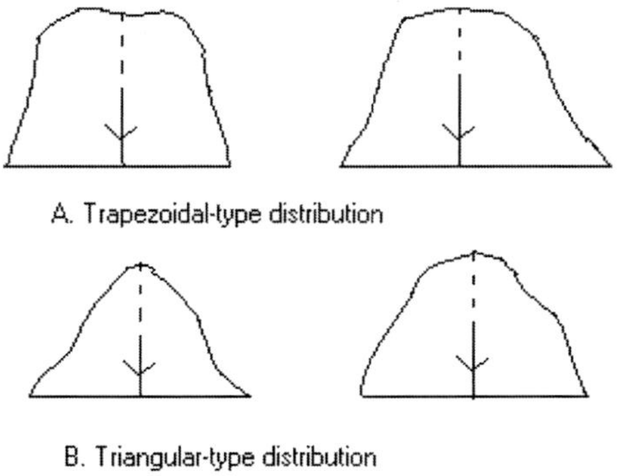

Figure 8.8: Characteristic water distribution pattern of sprinkler

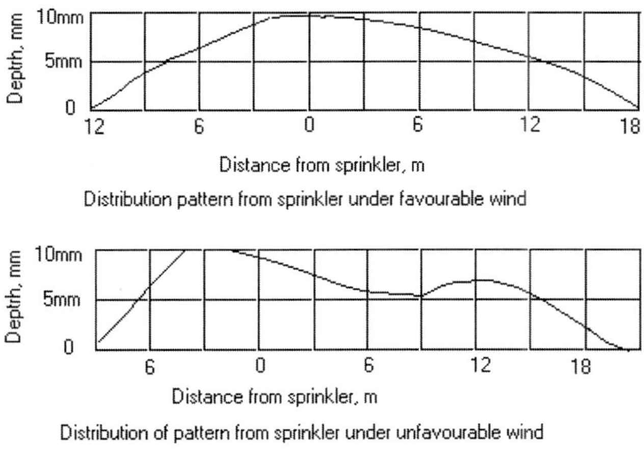

Figure 8.9: Probable distribution pattern of rotating sprinkler

The uniformity of water application may be expressed by a uniformity coefficient or to percentage. The uniformity coefficient may be acceptable when it is equal or higher than assumed value. From the Fig. 8.9 it is observed that depth of water is progressively diminishing towards away from the sprinkler. Therefore, to achieve acceptable distribution uniformity there must be overlapping of water application from the adjacent sprinklers. The degree of overlapping depends on the characteristics distribution pattern of the sprinkler and which in turn depends on sprinkler type, nozzle diameter, operating pressure and wind condition.

Usually less overlap is required for sprinklers that provide more or less trapezoidal distribution pattern than the triangular pattern. In general, the rectangular arrangement of sprinklers provide wider spacing between the sprinklers, the distance between the laterals is about 60-65% and 70-75% for triangular and trapezoidal respectively and the spacing between the sprinklers on the lateral is equivalent to about 40% of the wetted diameter are expected to produce an acceptable distribution uniformity. However, this recommendation may be considered as a guide only. The basis of spacing of sprinklers to be decided following the manufacturer's catalogs. It is better to conduct the test to know the exact distribution pattern of the sprinklers. Benami (1984) has suggested that spacing between the laterals to be shortening by 10% when the laterals are set at right angle to the direction of wind.

Field test for rotating head sprinklers

The field test is conducted to evaluate the uniformity of distribution from sprinklers for a given sprinkler and nozzle(s), pressure head, spacing and climatic condition. Several field tests are required with the same sprinklers under varying size of nozzles, pressure head and spacing to have the reference of sprinklers performances. The tests to be conducted at varying level of wind condition often the sprinklers are subjected to.

The rain gauges, which are usually the ordinary can of 1 litre are used to collect the water from the sprinklers. The wetted area around the sprinklers is divided in to squares of small area and at the center of each of it the cans are placed. The size of these small squares depends on the spacing of sprinklers or diameter of coverage. If the test is conducted in a crop field the cans to be raised over the crop height with the suitable support. The test runs of the sprinklers to be for sufficient time and approximately one-half of the planned time of irrigation. The spacing of cans are generally 2m in each direction for common rotating sprinklers spaced about 10m apart, 3m when sprinklers spacing are larger, and 1m for sprinkler of low flow-rate.

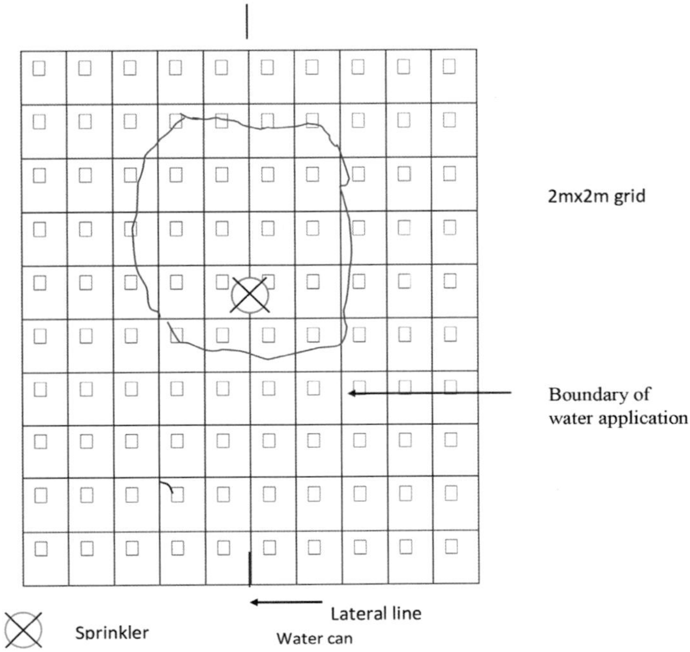

Figure 8.10: Water cans and sprinkler arrangement for testing of a single sprinkler

The lateral to be equipped with valve and pressure regulator to have close the sprinklers and varying the pressures. In a single sprinkler test for evaluating its pattern of application and uniformity around, it is to be placed in the center of the test area and the cans are placed in center of the square-grid pattern surrounding the sprinkler on all sides (Fig. 8.10). The boundary of the water application can be drawn by interpolation from the water received in the cans.

Uniformity coefficient

The degree of uniformity of application of water in the wetted area of a sprinkler determines how effectively the application is made. Pressure-nozzle size relation, sprinkler spacing and the wind condition influence the uniformity of application. The coefficient of uniformity may be computed by the depths of water caught in the cans placed at regular interval in the area of sprinkling. It may be expressed by the following equation as suggested by Christiansen (Michael, 1978):

$$C_u = 100\left(1.0 - \frac{\sum x}{mn}\right) \qquad (8.1)$$

where, C_u = uniformity coefficient in percent

m = average rate of water application, mm

n = total number of observation, n

x = numerical deviation of individual observation from the average application rate, mm.

The sprinklers are in use are few in number in each laterals at equal space and the laterals are placed in parallel in each other in set. Since the application of water from an individual sprinkler decreases outer wards, the application of sprinklers requires overlapping from every direction at certain extent for better application efficiency. For the purpose of test usually there will be overlapping of four sprinklers in rectangular arrangement. However, the square or triangular arrangement also can be made. In field practice, by operating the sprinklers and taking some probable spacing between the laterals (S_m) the application rates at different grid points in between the two opposite sprinklers may be observed. With the certain overlapping, the application rate may have maximum uniformity. Having this fixed, the spacing of the sprinklers on the laterals (S_l) to be varied and the uniformity of application to be examined. At certain overlapping there will be a best uniformity. The best spacing of sprinklers at certain pressure and other conditions may also to be determined by using the performance of the individual sprinkler. Assuming the sprinklers with certain spacing between the laterals S_m and along the laterals S_l at four corners of the rectangular arrangements, superimpose the identical pattern of each of the sprinklers so that the small squares around the boundaries are superimposed nicely to one upon another. The accumulated application of water at each superimposed square can be readily calculated. This will provide an approximation to the distribution of water application bounded by the sprinklers. A few trials may be required to arrive at the best approximation.

The loss of water in the form of evaporation during the process of sprinkling can be determined by measuring the discharge from the sprinkler head at the operating pressure. Pressure can be measured by using a pressure gauge fixed with a pitot tube. The pressure and discharge of the sprinkler head, on either side of the nozzles are measured before and after the test run. The pitot tube is inserted in to the nozzle to measure the pressure (Fig. 8.11). The discharge is measured by allowing the discharge of the sprinkler nozzle to a container by connecting through a flexible tube for a specified period. The difference between the discharge received in the container and the discharge recorded in the catch cans for the same specific time period denotes the evaporation loss of that period.

Courtesy: INCID (1998) *Courtesy:* Michael (1985)

Figure 8.11: Measurement of sprinkler nozzle pressure

The application rate of the sprinklers should be such that the applied water in the soil should disappear from the surface before the next application. Thus, the rate of application is always less or equal to infiltration rate of the soil. There should not be any runoff or movement of water over the ground. The movement of water over the ground indicates excess of water application. The infiltration rate in the soil is higher at the beginning and it decrease with the advancement of time and reaches to basic infiltration arte. The sprinkler operation is also takes place for few hours. Therefore, to ensure the instant infiltration of water, the infiltration rate after few hour of water application to be used in designing sprinkler operation.

Example 8.1 Determine the coefficient of uniformity in application of water from the following data as obtained in a field test on a square plot bounded by sprinklers. The depth of water is measured in millimeter.

○	14.3	13.6	13.0	13.8	14.3	○
14.0	13.9	12.9	14.7	15.4	15.1	14.8
14.5	13.6	12.7	15.6	12.9	14.0	14.7
14.1	14.1	13.1	14.6	13.6	13.9	14.0
14.2	13.9	11.8	12.3	14.4	14.8	13.9
○	14.9	14.2	13.5	14.6	13.0	○

3m × 3m grid

Solution: The computation of for coefficient of uniformity is shown in tabular form as below:

Observation	Frequency	Application rate x frequency	Numerical deviations	Frequency × deviations
15.6	1	15.6	2.005	2.005
15.4	1	15.4	1.855	1.855
15.1	1	15.1	1.555	1.555
14.9	1	14.9	1.355	1.355

14.8	2	29.6	1.255	2.510
14.7	2	29.4	1.155	2.310
14.6	2	29.2	1.055	2.110
14.5	1	14.5	0.955	0.955
14.4	1	14.4	0.855	0.855
14.3	2	28.6	0.755	1.510
14.2	2	28.4	0.655	1.310
14.1	2	28.2	0.555	1.110
14.0	3	42.0	0.455	1.365
13.9	4	39.6	0.355	1.420
13.8	1	13.8	0.255	0.225
13.6	3	40.8	0.055	0.165
13.5	1	13.5	0.045	0.045
13.1	1	13.1	0.445	0.445
13.0	2	26.0	0.545	1.090
12.9	2	25.8	0.645	1.290
12.7	1	12.7	0.845	0.845
12.3	1	12.3	1.245	1.245
11.8	1	11.8	1.745	1.745
	Σ=38	Σ=514.7, Mean=13.545		Σ=29.32

$$C_u = 100\left(1.0 - \frac{\sum x}{mn}\right)$$

$$= 100\left(1.0 - \frac{29.32}{514.7}\right)$$

$$= 100\,(1.0 - 0.056)$$

$$= 94.30\ \%$$

Design of Sprinkler Irrigation System

The objective of design of a sprinkler irrigation system is to achieve the maximum irrigation efficiency at low annual operation and maintenance cost. The design principles take the accounts of crop water requirements, soil types, topography, source of water, water quantity and quality, labour, economics and future scope for expansion. The steps of design are discussed below.

Inventory of the area

a. **Map of the area**: It is necessary to have a map of the area in scale so that the dimension of the area is understood as well as the probable lengths of mains and laterals of the sprinkler system. The map also shows the important objects around the field viz., roads, buildings, drainage channel, etc.

b. **Topography**: The map should have the contour lines in it at suitable contour interval to understand the undulation and slopes of the fields. The elevations of the important points like the location of the pump, water supply, elevation of ends of mains and laterals etc., to be delineated on the map.

c. **Climate:** The rainfall, temperature, evaporation, wind velocity, radiation intensity and humidity are required to know the consumptive use of the crop. The sprinkler system is designed on the basis of average of daily peak consumptive use rates in summer season assuming irrigation is done daily or alternate day.

d. **Water source-quantity, quality and period of availability:** The sprinkler system is designed to meet up the water requirements of selected crops at its maximum consumptive use as well the seasonal and annual requirements. The quality of water should be such that it should not have any corrosive effect to the equipments. The water used in the sprinkler system should be relatively clean and the level of suspended particles should be such that the sprinkler lines and nozzles do not get clogged. Off course, the type of crops and soil type determine the limit of chemicals permissible in irrigation water.

e. **Depth of irrigation:** The net depth of water may be calculated by using the following equation:

$$IR_n = \sum_{i=1}^{n} \frac{\left(M_{fci} - M_{bi}\right) A_s D_i}{100} \tag{8.2}$$

where, IR_n = net irrigation requirement, cm

M_{fci} = field capacity of the soil in the ith layer, percent

M_{bi} = moisture content in the ith layer at the time of irrigation, percent

A_s = apparent specific gravity of soil

D_i = depth of the ith soil layer in root zone depth, cm

n = number of soil layer in root zone depth.

The gross depth of water application may be calculated by dividing the net depth with the application efficiency.

f. **Crops grown:** Each crop has its own characteristics in respect of root zone depth, peak consumptive use rate and its time of occurrence. The depth of water application and availability of water at the critical periods are judged in respect to the crops.

g. **Irrigation interval:** The irrigation interval is the time period in days between two successive irrigations. It may be calculated by using the following formula:

Irrigation interval, I_i (days)

$$= \frac{Depth\,of\ irrigation\ (cm)}{Peak\ rate\,of\ daily\,consumptive\,use\ (cm/day)}$$

The irrigation interval is designed on the basis depth of water application. However, when the root system is at developing stage the irrigation interval in practice reduces and proportionately the depth of irrigation water.

h. **Application rate of water:** The discharges from the sprinklers to be determined based on the soil characteristics and land slope. The rate of application should not exceed the infiltration capacity of the soil. If it exceeds the infiltration capacity it will cause runoff resulting poor distribution of water, loss of water and soil erosion. The exact limiting value of the infiltration rate for particular field situation to be determined through experimentation. However, the values suggested in Table 8.7 for different soil conditions may be used when reliable data are lacking.

Table 8.7: Suggested maximum applications rates for sprinklers for average soil, slope and tilt

Soil texture and profile	0-5% slope	5-8% slope	8-12% slope	12-16% slope
	cm/h	cm/h	cm/h	cm/h
1. Coarse sandy soil to 2m	5.0	3.7	2.5	1.3
2. Coarse sandy soils over more compact soils	3.7	2.5	2.0	1.0
3. Light sandy loams to 2m	2.5	2.0	1.5	1.0
4. Light sandy loams over more compact soils	2.0	1.3	1.0	0.8
5. Silt loams to 2m	1.3	1.0	0.8	0.5
6. Silt loam over more compact soils	0.8	0.6	0.4	0.3
7. Heavy textured clays or clay loams	0.4	0.3	0.2	0.1

Table 8.8: Maximum spacing of sprinklers under windy condition

Sl. No.	Average wind speed	Spacing
1.	No wind	65% of the diameter of the water spread area of a sprinkler
2.	0-6.5 km/h	60% ,,
3.	6.5-13 km/h	50% ,,
4.	Above 13 km/h	30% ,,

Selection of sprinkler nozzles and spacing

Normally the sprinkler nozzle should be such that it gives the application rate equal or less than the infiltration rate of the soil. The sprinkler nozzle's catalog provided by the manufacturer shall include its model, size, diameter of throw, application rate and discharge.

The spacing of sprinklers depends on the diameter of throw and wind condition. The uniformity of water application in windy condition depends on the suitable overlapping of water spread area of sprinklers. In general, the overlapping increases with the increase in wind velocity. The Table 8.8 may be used as a guideline for designing the overlapping of sprinklers under different wind conditions. Table 8.9 & 8.10 are useful in determining soil moisture availability and depth of irrigation water.

Application rate of sprinkler (cm/h),

$$\frac{Depth\,of\ irrigation\ \ (cm)}{Peak\ rate\,of\ daily\,consumptive\,use\ \ (cm\,/\,day)} \qquad 8.3$$

Alternately, assuming the infiltration capacity of th

e soil as the maximum permissible rate of application of water,

$$q = \frac{S_l \times S_m \times I}{360} \qquad (8.4)$$

where, S_l = spacing of sprinklers along the laterals, m

S_m = spacing of laterals along the main, m

I = infiltration capacity of the soil, cm/h

Time needed for applying required depth of irrigation (hours),

$$T = \frac{depth\,of\ irrigation\,(cm)}{application\,rate\,of\ sprinkler\ nozzle\,(cm\,/\,h)}$$

Number of shifting of sprinkler system per day (n),

$$n = \frac{duration\,of\ pumping\ (h\,/\,day)}{T + time\ for\ each\,shifting\cdots(h)}$$

Area to be irrigated per day (ha/day), Area to be irrigated per shift, $A_2 = \frac{A_1}{n}$

Number of nozzle per shift $= \frac{A_2}{area\,covered\ per\ nozzle}$

Capacity of the sprinkler system

The capacity of the sprinkler system depends on the size of the area to be irrigated, irrigation interval, time of each irrigation and the gross depth of water application. Thus,

$$= \frac{A_2}{area\, covered\ per\ nozzle} \tag{8.5}$$

Where, Q = discharge capacity of the pump, l/s

A = area to be irrigated (the entire field), ha

d = net depth of water applied, cm

I_i = irrigation interval, days

H = time in hours per irrigation

E = water application efficiency, percent

Example 8.2 Calculate the sprinkler system capacity from the following data:

Number of laterals in the main = 2

Length of each lateral = 150m

Spacing of the sprinklers in the lateral = 12m

Spacing between the lateral lines = 16m

Water application rate = 1.0cm/h

Solution: Using the Eq.8.4, $q = \frac{S_l \times S_m \times I}{360}$

$$= 0.53l / s / sprinkler$$

Number of sprinkler in each lateral $= \frac{150m}{12m} = 12.5 \cong 13$

System capacity = discharge of each sprinkler x number of sprinklers

= 0.53l/s × 13 × 2

= 13.86l/s

Example 8.3 A farmer has a portable sprinkler system of one lateral of length 200m, sprinklers are spaced 12m on the lateral and spacing of lateral lines are 15m, the sprinklers sprays water at the rate of 1.5cm/h. The farmer likes to irrigate his 5ha wheat field with 6cm water. Assuming 9 hours working day, determine the days required to complete the irrigation and the discharge from each sprinkler. Allow 30 minutes for each shifting of the system.

Solution:

5 ha = 5 × 10000 = 50000m²

Let the size of the field = 250m × 200m

Time for 6cm application of water = $\frac{6cm}{1.5cm/h} = 4h$

Time for irrigation and shifting for each setting

= 4h + 0.5h = 4.5h

No. of shifting per day = $\frac{9}{4.5} = 2$

Each setting covers the entire width of the field and lengthwise 15m.

Therefore, number of shifting required = $\frac{250}{15} = 16.67 \cong 17$

No. of days required to irrigate the field = $\frac{17}{2} = 8.5$

The area covered by each sprinkler = 12m × 15m = 180m²

Volume of water discharges by each sprinkler in 1 hour

= 180m² × 1.5cm

= 2700 liters

Sprinkler discharge rate = $\frac{2700}{3600} = 0.75l/s$

Example 8.4 Fifteen sprinklers with twin nozzle of 5mm & 4mm diameter each with coefficient of discharge 0.96 are operating at 2.5kg/cm² pressure. The sprinkler spacing is 12m x 16m. the consumptive use rate for a particular crop is 6mm per day and irrigation interval is 10 days. Determine the (i) discharge of sprinkler, (ii) total capacity of the sprinkler system, and (iii) time of operation of sprinkler system at 75% efficiency (GATE, 1999).

Solution:

i) Discharge from the sprinkler =

$$C_d a\sqrt{2gh} = 0.96x\pi\left(\frac{d_1^2}{4} + \frac{d_2^2}{4}\right)\sqrt{2\times 9.81\times 25}$$

$$= 0.96x\pi\left(\frac{0.000025 + 0.000016}{4}\right)\times 22.147$$

$$= 0.0006846m^3/s = 2.466m^3/h = 0.685l/s = 2464l/h$$

ii) System capacity = 0.685l/s × 15 = 10.275l/s

iii) Application rate of sprinklers = $\frac{2.466m^3/h}{12m \times 16m} = 1.28cm/h$

Depth of each application = 6mm x 10 = 6cm

Time of operation at 75% efficiency $= \frac{6cm}{1.28cm/h \times 0.75} = 6.25h$

Example 8.5 Design the sprinkler system capacity for irrigating a 10ha field assuming the consumptive use rate as 5mm/day, maximum depth of application of water in one irrigation is 5cm, allowable leisure period within the irrigation interval is 2 days, efficiency of irrigation is 85%, and maximum operating period of the system in a day is 18 hours.

Solution:

Irrigation interval = $\frac{5cm}{0.5mm/day} = 10days$

Irrigation period = 10days - 2days = 8days

Using the Eq. 8.5, $Q = 2780\frac{A \times d}{I_i \times H \times E}$

$$= 2780\frac{10 \times 5}{8 \times 18 \times 85} = 11.36l/s$$

Example 8.6 Design the sprinkler irrigation system capacity for 15ha field under the following condition:

Soil = clay loam

Crop = cereals

Field slope = negligible

Duration of operation = 15h/day.

Efficiency of the system = 80%

Consumptive use = 0.5cm/day

Assume any other data if necessary.

Solution

From the Table 8.9 the average field capacity of the soil = 14cm/m

From Table 8.10 for cereal crops the effective root zone depth = 120cm

Available soil moisture $= \frac{14 \times 120}{100} = 16.8cm$

Let the irrigation is given at 50% depletion of available soil moisture.

Depth of water application = 16.8/2 = 8.4cm

Irrigation period = 8.4cm/0.5cm = 16.8days @ 17days

Using the Eq. 8.5, $Q = 2780 \frac{A \times d}{I_i \times H \times E}$

$$= 2780 \frac{15 \times 8.4}{17 \times 15 \times 80}$$

$$= 2780 \times 6.17 \times 10^{-3} = 17.1 l / s$$

Type of system and layout

The different type of sprinkler systems has been discussed earlier. The performances of the rotational head sprinklers are also described in Table 8.6. With these references, the sprinkler type and system may be determined. Next, the location of the pump and orientation of the mains and laterals are fixed based on the operating pressure, application rates, crop requirements and availability of labor.

The location of the pump is usually fixed with the location of source of water. In sprinkler irrigation the water is mostly obtained from tube well or open well source within the field. However, the river water source is also in use. If the situation permits, the point of highest elevation or middle of the field may be best choice for the location of the pump. This is because of using the advantage of elevation for distribution of water or to minimize the distances the water flows from the pump. The main always connected with the pump and its direction is determined by the location of source of water with the field. In case the existing underground pipelines are used, the portable pump unit is connected to the hydrants mounted on the pipe outlets. When the system is associated with permanent pumping unit and buried pipe lines, the pipe lines are usually run down the center of the field such that there will be little scope of hindrance to farm operations and movement of farm equipments. The mains of the sprinkler system follow the steepest slope and the laterals are at right angle thereto to have uniform pressures in the laterals and thereby the uniform rate of discharges. The design principle is thus putting the laterals on level surface. The variation of slope along the direction of lateral for a considerable length causes great difficulties for maintaining uniformity in application.

There may be many possible arrangements of mains, laterals and sprinklers. The best one should have the minimum cost of operation without compromising the continuous quality service. However, the choice is greatly influenced by the type and capacities of the sprinklers and their operating pressures. The availability of labour and the adverse conditions that to be encountered should get due consideration in selecting the arrangement.

Hydraulic design of sprinkler system

It is desired that the sprinkler irrigation system should maintain the uniformity in irrigation coverage at desired rate of application, the break-up of water drops should be such that it will cause minimum deterioration to soil structures and overall it runs efficiently by spending the minimum energy to cover the maximum area. The important hydraulic principles are given below.

Discharge of sprinkler nozzle: The discharge through the sprinkler nozzle may be computed by the following orifice formula as suggested by Torecelli (Michael, 1978):

$$q = CaV$$

$$= Ca\sqrt{2gh} \qquad (8.6)$$

Where, q = sprinkler nozzle discharge, m^3/s

a = cross-section area of nozzle, m^2

h = pressure head at the nozzle, m

g = acceleration due to gravity, m/s^2

C_d = Coefficient of discharge (C_d usually varies from 0.95 to 0.96).

Water spread area of sprinkler: The area of coverage by a rotating head sprinkler may be estimated by using the formula suggested by Cavazza (Michael, 1978):

$$R = 1.35\sqrt{dh} \qquad (8.7)$$

where, R = radius of the wetted area covered by the sprinkler, m

d = diameter of nozzle, mm

h = pressure head at the nozzle, m

The best coverage is attained when the sprinklers are set at an angle 30-32^0 from the horizontal. The most of the sprinklers are standardized at 30^0.

Break-up of jet: The proper break-up of jet is important for uniform coverage of area and to protect the soil structures on the surface. The low pressure than the standard causes to form large drops and higher throw and the drops fall with higher velocity. The high velocity drops have great adverse impact on deterioration of soil structures. High-pressure in nozzle breaks the jet at finer drops cannot have the desired throw. There is also scope of excess loss of water through evaporation. Therefore, there should be some sort of compromise among the distance of throw, uniformity of coverage, and effect on the soil surface. The following empirical formula suggested by Tanda (Michael, 1978), which provides the index of jet break-up:

$$P_d = \frac{h}{(10q)^{0.4}} \tag{8.8}$$

Where, P_d = index for jet break-up

h = pressure head at nozzle, m

q = sprinkler discharge, l/s

It is found that if P_d is greater than 2 the drop size is good; if 4, drop size is best; and if more than 4 the pressure is wasted.

Example 8.7 A sprinkler discharges at the rate of 0.75l/s. Determine the quality of spray and radius of spray if the diameter of nozzle is 3.96mm and operating at pressure 2.0kg/cm².

Solution:

Pressure head = 2.0kg/cm² = 20m of water

Using the Eq.6.8, break-up of jet, $P_d = \frac{h}{(10q)^{0.4}}$

$$= \frac{20}{(10 \times 0.75)^{0.4}}$$

$$= \frac{20}{(7.5)^{0.4}} = \frac{20}{2.23} = 8.93$$

The P_d value is more than 4. Therefore, the sprinkler operates under high pressure causing wasting of energy.

Using the Eq. 8.7, radius of spray, $R = 1.35\sqrt{dh}$

$$= 1.35\sqrt{3.96 \times 20} = 12.01m$$

Design of sprinkler laterals

The sprinkler laterals have equally spaced sprinklers along its length as if a pipe with evenly spaced multiple outlets. The flow decreases as it advances along the direction of flow and resulting the decreased pressure or the head loss in pipe-sections between successive outlets gradually towards the downstream.

Thus, the friction loss in a lateral pipe is much less than the pipe of same length if the total flow is carried to the entire length of the pipe. The Fig. 8.12 depicts the hydraulic grade line in a lateral over a level surface. It is assumed that the average operating pressure is located about 2/5th (for simplicity it can be assumed at the middle) of the lateral downstream from the inlet and at this point the head loss is 3/4th of the total head loss.

Let the head loss, H_f, along a lateral of diameter d and n number of sprinklers spaced at S_l on the lateral. Let the sprinkler at the farthest end (no.1) of the lateral discharges at the rate q_1 at pressure H_1. Therefore, the head loss in between the section of the sprinkler of farthest end (no.1) and next to that (no.2) is,

$$H_f(1-2) = J\frac{S_l}{100} \tag{8.9}$$

Where, J = head loss in 100m length of the aluminum or plastic lateral obtained from tables, nomographs or equations.

The pressure head at sprinkler 2 is,

$$H_2 = H_1 + H_f(1-2)$$

The discharges through the nozzle is,

$$q = C_d a\sqrt{2gH} \tag{8.10}$$

Where, q = sprinkler discharge, l/s

C_d = coefficient of discharge

a = cross-sectional area of sprinkler nozzle, m^2

H = operating pressure head, m

Assuming the C_d constant at all discharges from the sprinkler nozzle,

$$\frac{q_2}{q_1} = \frac{C_d a\sqrt{2gH_2}}{C_d a\sqrt{2gH_1}} = \sqrt{\frac{H_2}{H_1}} = \sqrt{\frac{H_1 + H_f(1-2)}{H_1}} \tag{8.11}$$

$$\text{or, } q_2 = q_1\sqrt{1+\frac{H_f(1-2)}{H_1}} \tag{8.12}$$

Similarly, $H_3 = H_2 + H_f(2-3) = H_1 + H_f(1-2) + H_f(2-3)$ (8.13)

$$q_3 = q_1\sqrt{1+\frac{H_f(1-2)+H_f(2-3)}{H_1}} \tag{8.14}$$

Step by step this procedure may be repeated till the sprinkler in m upstream (m<n) when $H_f = (H_m - H_1) \geq H_v$, where H_v is the maximum permissible

variation of pressure. When, $H_f = (H_n - H_1) \leq H$ the n number of sprinklers can be safely used. If the situation is so, the length of the lateral may be shortened or the diameter may be increased to accommodate the pressure variation within the permissible limit. In case, when H_n and H_1 are known, the q_n and q_1 are also known. The average discharge through the sprinklers may be calculated by dividing the total flow in lateral (Q_n) by n. Thus the average pressure in the lateral may be determined by,

$$\frac{q_a}{q_1} = \sqrt{\frac{H_a}{H_1}} \tag{8.15}$$

or

$$H_a = H_1 \left(\frac{q_a}{q_1} \right)^2 \tag{8.16}$$

Therefore, the sprinkler discharges q_a at pressure H_a may be selected from the catalogs provided by manufacturers of sprinkler.

The local head loss due to aluminum quick-couplers with no outlets is considered negligible. However, when the pipe lengths do not match with the sprinkler spacing, the quick-couplers are installed in between the sprinklers; the head loss cannot be ignored. Head loss due to 'saddles' used in plastic laterals is also negligible. However, the head loss in pressure regulating, low flow rate sprinklers or 'spitters' is not negligible. The local head loss usually remains 6-7% and do not exceed 10%.

The procedure of calculating the head loss in lateral as described in Eq. 8.9-8.16 is tedious and time consuming. Christiansen (Benami & Often, 1984) proposed a simple method, which calculates the head loss in lateral by assuming the same discharge through all the sprinklers. The computation of head loss by this way also gives most approximate result.

Thus, $$\frac{Q_n}{n} = q_1 = q_2 = \cdots = q_n = q_a \tag{8.17}$$

The head lost in a pipe with the diameter D, discharge Q and length L can be determined by the equation (following Darcey-Weisbach),

$$H_f = \frac{KLQ^r}{D^{2r+1}} \tag{8.18}$$

The head loss in pipe section from first to second sprinkler from the tail end,

$$H_f(1-2) = \frac{KS_l q^r}{D^{2r+1}} \tag{8.19}$$

The head loss in pipe section between sprinklers 2 to 3 from the tail end,

$$H_f(2-3)=\frac{KS_l(2q)^r}{D^{2r+1}} \tag{8.20}$$

Similarly, the head loss in between the sprinklers n to lateral inlet from the tail end,

$$H_f(n-inlet)=\frac{KS_l(nq)^r}{D^{2r+1}} \tag{8.21}$$

In a level field the sum of the losses in the lateral,

$$H_f=\frac{KS_l(q)^r}{D^{2r+1}}\left(1+2^r+\cdots+n^r\right)=\frac{KS_l(q)^r}{D^{2r+1}}\sum_{i=1}^{n}i^r \tag{8.22}$$

Putting, $q_a=\frac{Q_n}{n}$, and , $S_l=\frac{L}{n}$ $(q=q_a)$

$$H_f=\frac{KLQ_n^r}{D^{2r+1}}\cdot\frac{1}{n^{r+1}}\sum_{i=1}^{n}i^r=\frac{KLQ_n^r}{D^{2r+1}}\cdot F \tag{8.23}$$

where, $F=\frac{1}{n^{r+1}}\sum_{i=1}^{n}i^r$

Thus, it appears that the head loss along a multiple outlet pipe is the product of a coefficient to the head loss of the pipe if the entire flow passes through the outlet at the end of the pipe. The coefficient F depends on the number of outlets, n, along the lateral, the location of the first outlet with respect to inlet and the type of materials used for making the laterals.

The value of the coefficients are described in Table 8.11 & 8.12.

The value of coefficient F also can be calculated by using the Christiansen (1942) equation as below.

$$F=\frac{1}{b+1}+\frac{1}{2N}+\frac{(b-1)^{0.5}}{6N^2} \tag{8.24}$$

Where, b =1.852

N = number of sprinklers

Example 8.8 Determine the multiplication factor F for friction losses in a sprinkler lateral of 5 outlets.

Solution

No. of outlets in the lateral, n = 5. Considering Darcey-Weisbach equation where power of q = 2 i.e., r = 2

$$F = \frac{1}{n^{r+1}} \sum_{i=1}^{n} i^r = \frac{1}{5^{2+1}} \left[1 + 2^2 + 3^2 + 4^2 + 5^2\right]$$

$$= \frac{1}{125} \times 55 = 0.44$$

F_1 is to be used when the distance from the lateral inlet to the first outlet is S_1 meters.

F_2 is to be used when the first outlet is near the lateral inlet.

F_3 is to be used when the distance from the lateral inlet to the first outlet is $S_1/2$ meters.

The design capacity of the sprinkler is based on average operating pressure. If the friction head loss and the average operating pressure are H_f and H_a respectively, the operating head at the inlet H_n and H_a can be expressed as the following,

$H_a = H_1 + 1/4 H_f + H_r$ (8.25)

$H_n = H_a + 3/4 H_f + H_r$ (8.26)

where, H_1 = pressure at the first sprinkler from the tail end

H_r = riser height

If the lateral is placed over the uniform sloping land the Eq.6.25 is modified as,

$$H_n = H_a + 3/4 H_f + H_r \pm \frac{\Delta Z}{2} \tag{8.27}$$

where, ΔZ = the elevation difference between the ends of the lateral

The positive value of elevation difference is considered when the lateral runs up slope and the negative when down slope. The half of elevation difference is taken as because the average pressure occurs almost at the middle length of the lateral.

Pressure variation limit

In designing the lateral the maximum pressure variation should not exceed 20 percent and the discharges 10 percent. However, in practice the pressure variation occasionally exceeds to the extent of 30 percent. When the pressure variation exceeds the limit this may be adjusted by increasing the diameter of the lateral or by shortening its length.

In a portable sprinkler irrigation system the irrigation starts from the last setting on the sub main. As the lateral advances in the successive setting along the sub main, the lateral inlet pressure, H_n, gradually increases. This causes to variation of pressure exceeding the allowable 20 percent limit. This problem may be overcome by using the take-off valves at each lateral and controlling it manually. However, this is become impractical when a few laterals are operated simultaneously along a sub main. Therefore, instead of the individual lateral, the set of laterals operated simultaneously along the sub main may be considered for pressure variation. The pressure variation if exceeds the 20 percent between the sub mains or manifolds that can be controlled by using the regulator at the inlet. The diameter of pipes of lateral and sub mains may be selected accordingly.

In the solid set system the pressure variation between the laterals may be allowed to exceed 20 percent if regulators are used to inlet of the laterals. This method provides the advantage of using smaller diameter pipes for laterals and sub mains. However, it requires higher energy at the water source.

In practice, the size of the lateral is generally selected by trial and error procedure taking in to consideration the cost of pipe and the friction losses. Some useful charts and tables are referred to estimate the friction losses for the given discharge, size and length of the lateral and size of the nozzles. The following steps are followed:

Step 1. Select a reasonable diameter of the pipe.

Step 2. Compute the friction loss of the pipe for the given discharge to flow the entire length of pipe without sprinklers/nozzles.

Step 3. Correct the friction loss found in step 2 by the multiplication factor F determined from Table 8.11-12 corresponding to number of sprinklers and location of the first sprinkler with respect to inlet.

Step 4. The friction loss in step 3 is adjusted to $\pm\Delta Z/2$ if the laterals goes uphill or downhill (positive for uphill and negative for downhill).

Step 5. Verify whether the head loss computed in step 4 is within the pressure variation of 20 percent or not. The diameter of the pipe is acceptable if the variation is within this limit. If excess low head loss is found a lower diameter pipe may be tried to save the cost of pipe. If the pressure variation exceeds the limit, the next larger diameter pipe may be tested with the repetition of step 1 to 5 to arrive at the proper selection.

It may be noted that usually the local head losses are ignored; however, when the losses are considerable that should be added in computing the friction loss in the lateral pipe. The friction loss in lateral pipe may also be calculated by Hazen-William (for plastic pipe) and Scobey's equation (for aluminum pipe).

Example 8.9 Compute the head loss of a lateral of length 320m, diameter 12.5cm aluminum pipe with couplers and made of 12m sections, sprinklers spaced 16m and discharge 1.25l/s, and the first sprinkler is at one sprinkler distance from the inlet.

Solution

$$\text{No. of sprinklers on the lateral} = \frac{320m}{16m} = 20$$

Total flow through the lateral $= 20 \times 1.25 = 25l/s$

From **Appendix H1** the friction loss in a 12.5cm aluminum pipe for a flow of 25l/s in it $= 3.28m/100m$

In **Appendix H (Table H-1)** it is advised in the footnote that 3% of the friction loss to be deducted for using 12m sections. From Table 8.11 the value of $F = 0.359$ when the first sprinkler is set at one sprinkler distance from the main.

Using Hazen-William equation (Eq. 4.7),

$$J = H_f = \frac{h_f \times 100}{L} = K\left(\frac{Q}{C}\right)D^{1.852} \text{ for 100m pipe of continuous flow,}$$

$$h_f = 1.212 \times 10^{12}\left(\frac{25}{150}\right) \times 125^{-4.87} = 2.69m$$

Using the Table H-1the friction loss,

$$h_f = \frac{3.28}{100} \times 320 \times \frac{97}{100} \times 0.359$$

$$= 3.65m$$

Using Christiansen (1942) (Eq.8.4), the multiplication factor

$$F = \frac{1}{b+1} + \frac{1}{2N} + \frac{(b-1)^{0.5}}{6N^2}$$

Where, b = 1.852, N= 20,

$$or,\ F = \frac{1}{2.852} + \frac{1}{40} + \frac{0.852^{0.5}}{6 \times 400} = 0.35 + 0.025 + 3.85 \times 10^{-4} = 0.375$$

Therefore, the head loss $h_f = 2.69 \times 0.375 \times \dfrac{320}{100} = 3.23m$

Example 8.10 Determine the diameter of the lateral from the following conditions:

Discharge of the sprinkler = 0.5l/s

Effective diameter of spray of the sprinkler along the lateral = 16m

Length of the field along the lateral line = 220m

Average operating pressure of the sprinkler = 21m

Pipes are available in the market = 7.5cm, 10.0cm, 12.0cm.

Solution:

No. of sprinklers required = $\dfrac{220}{16} = 13.75 \cong 14$

Total flow through the lateral $= 0.5 \times 14 = 7.0l/s$

The correction factor F for 14 sprinklers is found 0.370 from Table 8.11.From

Appendix H (Table H-1) the friction losses are computed as below for 7.0l/s discharge.

Diameter of pipe, cm	Friction loss for 100m length of lateral (H_1), m	Friction loss for 220m length of lateral and adjusted by the correction factor F $\left(H_f = H_1 \times F \times \dfrac{220}{100}\right)$, m
7.5	3.93	3.2
10.0	0.90	0.73

The pressure variation of both the pipes is within the permissible limit of 20 percent. Therefore, the low diameter 7.5cm should be selected.

Example 8.11 A level field of 300m × 300m is to be irrigated by portable lateral set made of aluminum pipes and fed from the sub main run through the center of the field. The sprinkler discharges 0.4l/s at average operating head of 28m and spaced 15m apart along the lateral. The first sprinkler is located at 7.5m from the lateral inlet. Design the diameter of the laterals.

Solution:

Number of sprinklers in a lateral $= \dfrac{150}{15} = 10$

Length of lateral pipe = 150-15/2 = 142.5m

Discharge through the lateral = 0.4l/s × 10 = 4l/s

For 10 outlet lateral pipe with first lateral located at half sprinkler spacing form the inlet (Table 8.11), F= 0.353

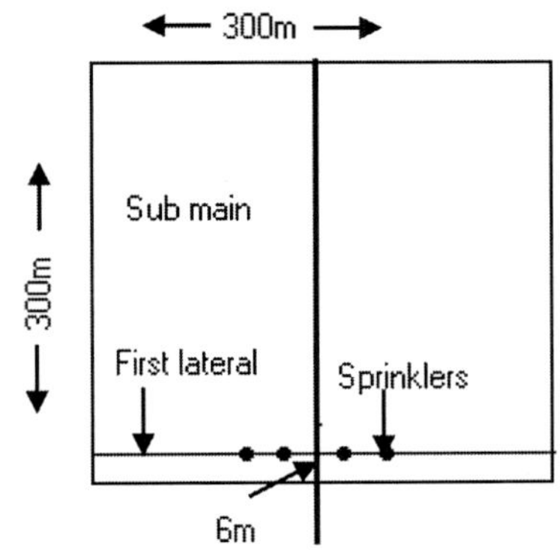

Maximum allowable head loss in the field $= 28 \times \frac{20}{100} = 5.6m$

Let the diameter of the lateral be 7.5cm.

The head loss in the lateral of 7.5cm diameter with 4.0ls discharge (Appendix H, Table H-3) =1.584m/100m

Head loss in the lateral $= \frac{1.584}{100} \times \frac{142.5}{100} \times 0.353 = 0.8m$

The head remaining may be lost in the sub main = 5.6-0.8= 4.8m

The lateral diameter of 7.5 cm is accepted.

Example 8.12 A level sub main is placed along the center of the rectangular field of 84m x 84m feeds the plastic laterals laid to both sides. There is a land slope of 5 percent across the direction of sub main. The sprinklers are spaced 6m and discharges 0.185l/s at operating pressure head of 28m. The first sprinkler is located at 3m from the sprinkler inlet. Determine the diameter of the laterals. Pipes are available 25mm, 31.8mm, 50.8mm and 62.5mm.

Solution

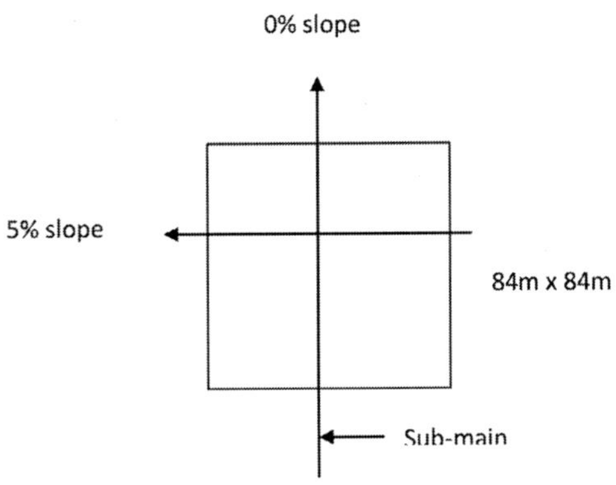

The maximum allowable pressure variation (20%) = 28 × 0.2 = 5.6m

The number of sprinklers on each lateral $= \frac{84}{2\times 6} = 7$

The discharge in the lateral, $q_n = 0.185 \times 7 = 1.295 l/s$

The value of F = 0.363

Length of the lateral $= \frac{84}{2} - 3 = 39m$

Let the lateral be tried for 25mm diameter for both the side.

Friction loss (Appendix H, Table H-4), $H_f = \frac{11.33}{100} \times 39 \times 0.363 = 1.61m$

For left side:

$$H_n = H_a + 3/4H_f + H_r \pm \frac{\Delta Z}{2}$$

$$= 28 + 3/4 \times 1.61 + 0.0 - \frac{1.95}{2} = 28.23m$$

$$H_1 = 28.23 - 1.61 + 1.95 = 28.57m$$

Pressure difference, $\Delta h = 28.57 - 28.23 = 0.34m$

For right side:

$$H_n = 28 + 3/4 \times 1.61 + 0.0 + \frac{1.95}{2} = 30.18m$$

$$H_1 = H_n - H_f - \Delta Z$$

$$= 30.18 - 1.61 - 1.95 = 26.62m$$

$$\Delta h = H_n - H_1 = 30.18 - 26.62 = 3.56m$$

In both the laterals the head loss is within the permissible limit. Therefore, the diameter of lateral 25mm is acceptable.

Maximum number of laterals operating simultaneously

To determine the maximum number of laterals operating simultaneously on a sub main (or manifold) it is necessary to know the irrigation water requirements of crops, irrigation intervals of various crops, the length of the sub mains or manifolds, the spacing of laterals, the number of daily irrigation and the daily operating hours. The details of this may be described by the following example.

Example 8.13 A sprinkler irrigation system is to be designed to irrigate a field in which three crops are grown in three successive seasons. The sprinklers are of 3.96mmx3.2mm size, operate at 28m average pressure head, and discharge 0.45l/s with the application rate of 10mm/h. Hand-moved aluminum laterals spaced 12m are used in a sub main of length 300m placed along the centre of the field. The irrigation continued 15 hours daily. The field capacity and wilting point of the soil are 35cm/m and 15cm/m respectively. Assuming the system efficiency 75%, determine the number of laterals that operate simultaneously. The additional data of soil and crops are given below.

Crop	Root zone depth, m	Consumptive use of crop, cm/day	Irrigation given at depletion of available soil moisture, percent
A	0.8	0.5	50
B	1.0	0.5	50
C	1.2	0.5	50

Solution

For crop A:

Total number of laterals $= \frac{300}{12} \times 2 = 50$

Irrigation water requirement

$$= (FC - PWP) \cdot D \cdot d_m = (35 - 15) \times 0.8 \times 0.5 = 8.0cm$$

Irrigation interval $= \frac{8.0cm}{0.5cm} = 16 days$

Assuming 2 days required for maintenance of the system and leisure period of the operator, the net irrigation days (irrigation period) in the irrigation period = 14 days.

Gross irrigation requirement $= \frac{8.0cm}{0.75} = 11.44cm$

Duration of an irrigation $= \frac{11.44cm}{10.0mm/h} = 11.44h$

No. of laterals required daily $= \frac{50}{14} = 3.57$

No. of daily application $= \frac{18}{11.44} = 1.57$

For crop B:

Irrigation water requirement $= (35 - 15) \times 1.0 \times 0.5 = 10.0cm$

Irrigation interval $= \frac{10.0cm}{0.5cm} = 20 days$

Assuming 3 days required for maintenance of the system and leisure period of the operator, the irrigation period = 17 days.

Gross irrigation requirement $= \frac{10.0cm}{0.75} = 13.33cm$

Duration of an irrigation $= \dfrac{13.33cm}{10.0mm/h} = 13.33h$

No. of laterals required daily $= \dfrac{50}{17} = 2.94$

No. of daily application $= \dfrac{18}{13.33} = 1.35$

For crop C:

Irrigation water requirement $= (35-15)\times 1.2 \times 0.5 = 12.0cm$

Irrigation interval $= \dfrac{12.0cm}{0.5cm} = 24.0days$

Assuming 4 days required for maintenance of the system and leisure period of the operator, the irrigation period = 21 days.

Gross irrigation requirement $= \dfrac{12.0cm}{0.75} = 16.0cm$

Duration of an irrigation $= \dfrac{16.0cm}{10.0mm/h} = 16.0h$

No. of laterals required daily $= \dfrac{50}{21} = 2.38$

No. of daily application $= \dfrac{18}{16.0} = 1.13$

The performances in crop A,B&C are summarized below in table form.

Crop	NWR, mm	Irri. Interval, days	Irri period, days	GWR, mm	Duration of application, h	No. of daily application	No. of laterals/ day	No. of laterals/ application
A	80	16	14	114.4	11.44	1.57	3.57*	2.2*
B	100	20	17	133.3	13.33	1.35	2.94	2.19
C	120	24	21	160	16.0	1.13	2.38	2.10

It appears from the above table that the maximum number of laterals in any crop is $2.27 \approx 3$. The sub main should be designed on the basis of 3 laterals.

Design of a sub main

The sub main has to convey the required discharge at desired pressure to all the laterals in it. The selection should be based on the consideration of cost of the pipes and cost of energy to operate the system. The length of the sub mains is usually considerable and the discharge gradually reduces towards the downstream, therefore, there is scope for selection of low diameter pipes from the upstream to downstream of the sub main. To arrive at the most economic diameters of the sub mains a step-by-step calculation is followed. An example is presented below.

Example 8.14. An aluminum sub main runs through the center of a field 288mx288m field, having a downward slope of 2.0 percent and buried 0.5m. The sub mains have 24 take-offs, spaced 12m and the total number of lateral settings are 48. Four level hand moved aluminium laterals are to be operated simultaneously along the sub main.The local head loss in a take-off is 1.5m. The selected sprinklers are 3.96mm × 3.2mm, spaced 12m × 12m, and apply water 0.51l/s at 35m head. A net pressure head of 39m is available at the connection to the main. Determine the diameter of the sub mains.

Solution

The schematic diagram of the field and the lateral setting at the farthest end are shown in Fig. 8.13

The design of the sub main is started at the setting to the farthest end 4′ (Fig. 8.13). The design pressure is required at the farthest sprinkler is 35m. The operating head of the sprinklers are required to overcome the local head loss (1.5m) and elevation difference due to buried sub main (0.5m) = 35+1.5+0.5=37m

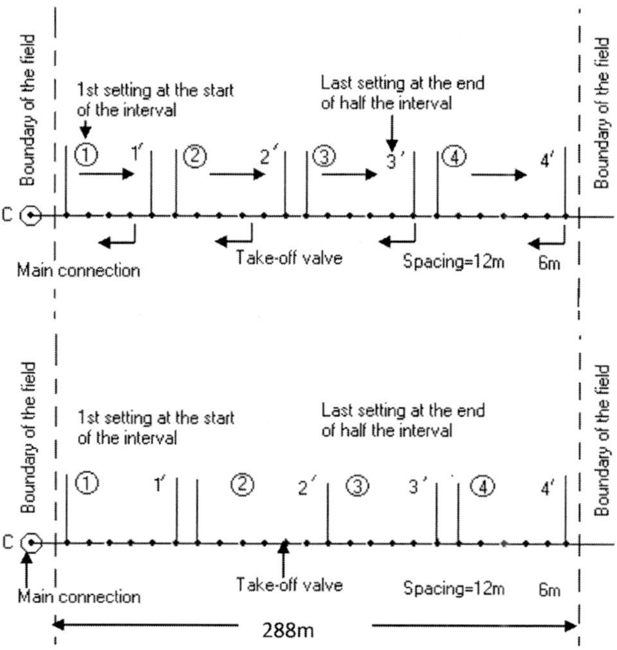

Figure 8.13: Sequence of Operation of lateration along the sub main in Example 8.14

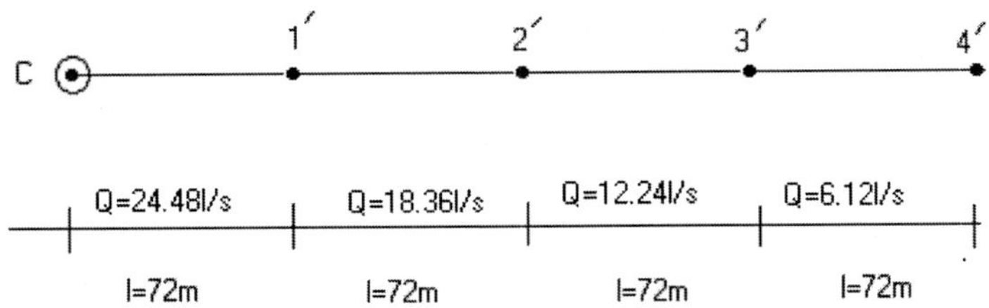

Figure 8.14: Diagram of discharge and lengths

Sub main section	Discharge, 1/s	Pipe diameter (D), mm				
		75	100	125	150	175
4'-3'	6.12	0.87	0.18	-	-	-
3 '-2'	12.24	3.53	0.83	0.26	-	-
2'-!'	18.36	7.60	1.81	0.59		
r-c	24.48	-	301.	1.03		

The elevation gain along the section $= \frac{288}{4} x \frac{2}{100} = 1.44m$

The number sprinklers in each lateral $= \frac{288}{2x12} = 12$

The discharge through the lateral $= 0.51x12 = 6.12l/s$

The value of F=0.369 [assuming 1st sprinkler is ½ sprinkler from the sub main and using Table 8.11 for 6 junctions of laterals in a section]..

The Table 8.13 shows the friction loss (h_f) for a number of selected diameters along each section of the sub main (72m) for the constant discharge following **Appendix H (Table H-3).**

The design starts at 4′, the farthest end of downstream. The adjusted pressure head is required 37m. A pipe of diameter 75mm selected initially for section 4′-3′ provides the head loss of 0.83m (Table 8.13). Therefore, head at the inlet with the adjustment of elevation difference,

$$h_n = 37 + 0.87 - 1.44 = 36.43m$$

The computed pressure head, $h_4 = 36.43m \prec 37m$. Therefore, pressure at this point may be raised by 0.57m. This could be done by using low diameter pipe. However, in sub main pipes less than 75mm is not commonly used. Therefore, for this section the 75mm diameter pipe is selected

For the next upstream section (3′-2′), $h_3 = 37m$

Assuming D = 100mm; h_f = 0.83m and elevation difference, $\Delta Z = 1.44mm$

$$h_n = 37 + 0.83 - 1.44 = 36.39m$$

The computed pressure head h_3= 36.39m<37m. Thus, head should be raised by 0.61m (37-36.39=0.61m). This could be done by using partially the 75mm pipe. The head loss in the pipes are required = 0.83+0.61=1.44m. Let x be the length of 100mm diameter pipe and (72-x) the length of 75mm pipe. Therefore,

$$h_{f(100mm)}(x)+h_{f(75mm)}(72-x)=1.44m$$

$$0.83/72(x) +3.53/72(72-x) =1.44$$

$$\text{or, } 0.0115277x+3.53-0.0490277x =1.44$$

$$\text{or, } 0.0375x=2.09$$

$$x = 55.73m$$

Thus, for the third section D = 100mm of 55.73m and D = 75mm for 16.27m.

For the next upstream section (2′-1′), $h_2 = 37m$

Assuming D = 100mm; h_f = 1.81m and elevation difference, $\Delta Z = 1.44mm$

$$h_n = 37 + 1.81 - 1.44 = 37.37m$$

The computed pressure head h_2= 37.37m>37m. Thus, head should be lowered by 0.37m (37.37-37=0.37m). This could be done by using partially the 125mm pipe. The head loss in the pipes are required = 1.81-0.37 =1.44m. Let x be the length of 100mm diameter pipe and (72-x) the length of 125mm pipe. Therefore,

$$h_{f(100mm)}(x)+h_{f(125mm)}(72-x) =1.44m$$

$$1.81/72(x)+0.59/72(72-x) =1.44$$

$$\text{or, } 0.025138x +0.59 -8.194\times10^{-3}x = 1.44$$

$$\text{or, } 0.016944x = 0.85$$

\x = 50.16m

\D = 100mm along 50.16m & D = 125mm along 121.84m for the section 2’-1’.

For the next upstream section (1′-C), $h_1 = 37m$

Assuming D = 125mm; h_f = 1.03m and elevation difference, $\Delta Z = 1.44mm$

$$h_n = 37 + 1.03 - 1.44 = 36.59m$$

The computed pressure head h_1 = 36.59m < 37m. Thus, head should be raised by 0.41m (37-36.59=0.41m). This can be done by using partially the 100mm pipe. The head loss in the pipes are required =1.44m. Let x be the length of 125mm diameter pipe and (72-x) the length of 100mm pipe. Therefore,

$H_{f(125mm)}(x)+h_{f(100mm)}(72\text{-}x) = 1.44m$

$1.03/72(x)+3.10/72(72\text{-}x) = 1.44$

or, $0.014305x + 3.01 - 0.04305x = 1.44$

or, $0.02875x = 1.57$

\x = 54.60m

$\therefore D = 125mm$ along 54.35m & $D = 100mm$ along 17.40m for the section 1′.-C.

Design of a manifold

The manifold is the sub-main in which all the laterals in it operate simultaneously. Therefore, its design is similar to a lateral. The design is expressed by an example.

Example 8.15 A level field of 144mx144m to be designed by solid-set sprinkler irrigation system. The manifold runs through the center of the field. The laterals are placed on both sides at 12m on the manifold and the sprinklers are also at 12m intervals on the laterals. The laterals operate simultaneously to irrigate the entire field. The selected sprinkler operates at pressure head of 20m at the rate of 500l/h. The local head loss in both the laterals and manifolds are 10 percent. The recommended maximum pressure variation is 20 percent. Design the pipe diameters (PVC).

Solution

The lateral:

The maximum allowable pressure head variation

The number of laterals $= \frac{144}{12} = 12$

The number of sprinkler in a lateral $= \frac{144}{2\times12} = 6$

$\therefore F = 0.369$ (1st sprinkler is ½ sprinkler from the manifold, Table 8.11)

Lateral inflow, $q_l = 500l/h \times 6 = 3000l/h = 0.83l/s$

Let the lateral pipe be 25mm. For 25mm diameter and with the flow of 0.83l/s the friction loss is 6.15m/100m (Appendix H, Table H-4).

Friction loss in the lateral, $h_f = 6.15 \times \frac{(72-6)}{100} \times 0.369 \times 1.1 = 1.18m \prec 4m$

Therefore, 25mm pipe can be safely used.

The inlet pressure head, $h_n = 20 + 3/4 \times 1.18 = 20.89m$

The pressure head at the downstream end, $h_l = 20.89 - 1.18 = 19.71m$

The manifold: No. of lateral junctions $= \frac{144}{12} = 12$

$\therefore F = 0.349$

Length of pipe $= 144 - \frac{12}{2} = 138m$

Manifold inflow, $Q_n = 12 \times 2 \times 3000 l/h = 72000 l/h = 20 l/s$

For 100mm pipe, friction loss $= 4.48m/100m$

$\therefore h_f = 1.1 \times 4.48 \times \frac{138}{100} \times 0.349 = 2.37m$

Pressure at the inlet of manifold $= 20.89 + \frac{3}{4} \times 2.37 = 22.67m$

The maximum pressure head in the field is in the first sprinkler of first lateral and minimum at the last sprinkler of last lateral. The pressure at the first sprinkler of first lateral is approximately 22.67m

The pressure head at the inlet to the last sprinkler $= 22.67 - 2.37 = 20.3m$

The pressure head at the last sprinkler of last lateral $= 20.30 - 1.18 = 19.12m$

The maximum variation of head in the field $= 22.67 - 19.12 = 3.55m$

The 3.55m is less than 4.0m and also very close to it. The design of pipes is satisfactory.

However, considering the various local head losses an estimated head of 25m at the connection to the main may be accepted to be on the safe side.

Design of Main Line

Main pipelines are responsible to convey the water to the various points in the irrigated fields. Usually the entire area under the consideration is not irrigated simultaneously but a few fields or sub plots are done. Therefore, to design the mains, the sequence of sub mains or manifolds to be examined with the support of schematic diagrams of each set of arrangements to arrive at the appropriate selection of sequence. The diagram includes the discharge, pipe lengths, elevation and pressure head required at each main and sub main. The selection of diameter starts from the downstream and gradually approaches to the upstream water source.

Example 8.16 A level field of 2 plots each consists of 6 sub-plots to be designed for solid-set sprinkler system The sub plots are similar to the fields designed in previous Example (Fig. 8.15). The designed discharge at the sub main inlet, $Q_n = 20 l/s$ and pressure head, $h_n = 25.0m$. The irrigation interval is 3 days and 4 sub-plots are irrigated simultaneously in a day. The available pressure head at the water source is 45m. Assuming local head loss 10% of the longitudinal head loss, design the PVC pipes.

Solution

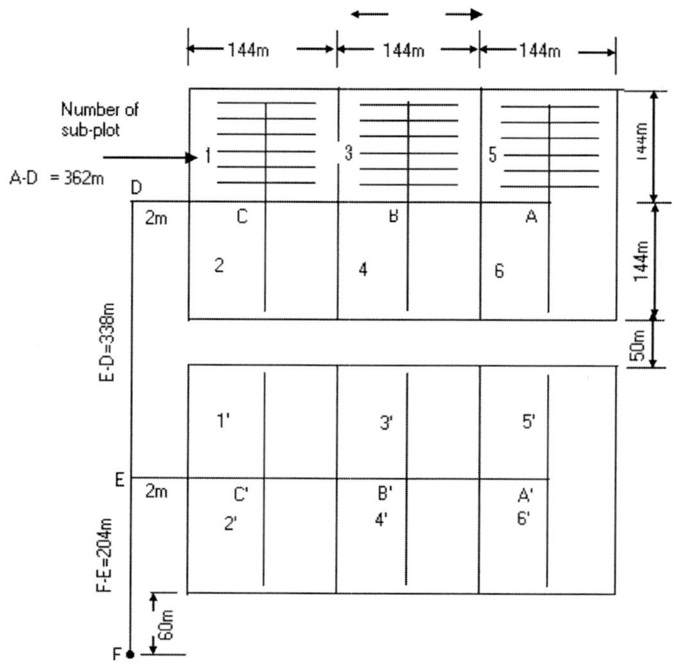

Figure 8.15: Plan of plots of the solid-set sprinkler system

Solution: As stated in the problem, 4 sub-plot to be irrigated in each day. There may be a few sequences of irrigation. Let the sequence of irrigation are 2,6,2'6'; 1,5,1'& 5' & 3,4,3' & 4' for the first, second and third day of irrigation respectively.

As shown in Fig. 8.16 there are two cases A & B. The case A shows the discharges for the first two days and the case B shows the same for the third day. These two cases may be treated separately to arrive at the recommended diameter of the mains.

The friction head losses for selected diameters along the pipe-sections at constant discharge are presented in Table 8.14.

Plot I

Case A: Design of main A-C-D-E-F

Using the Table 8.14 & Fig. 8.16 the diameter for AC, CD & DE are selected 125,150 & 150mm respectively with the head loss of 16.23m. This makes a resulting pressure head of 25 + 16.23 = 41.23m at E. Thus, the available head loss in EF = 45 – 41.23 = 3.77m. The Table 8.14 indicates that with 80l/s discharge there is scope for combination of 200mm & 250mm pipes in EF. Let x be the diameter of 200 mm pipe.

Therefore,

$$\frac{5.64}{74}\cdot x+\frac{1.89}{204}(204-x)=3.77$$

$$\text{or, }\left(\frac{5.64}{74}-\frac{1.89}{204}\right)x=3.77-\frac{1.89}{204}x204$$

$$\text{or, } 0.018336x=1.88$$

$$\therefore x=102.53m\approx 102m$$

Length of 250mm pipe = 204 – 102 = 102m

Summary of pipe selection:

Section	Length of pipe, m					
	75mm	100mm	125mm	150mm	200mm	250mm
AC			288			
CD				74		
DE				338		
EF					102	102
			288	412	102	102

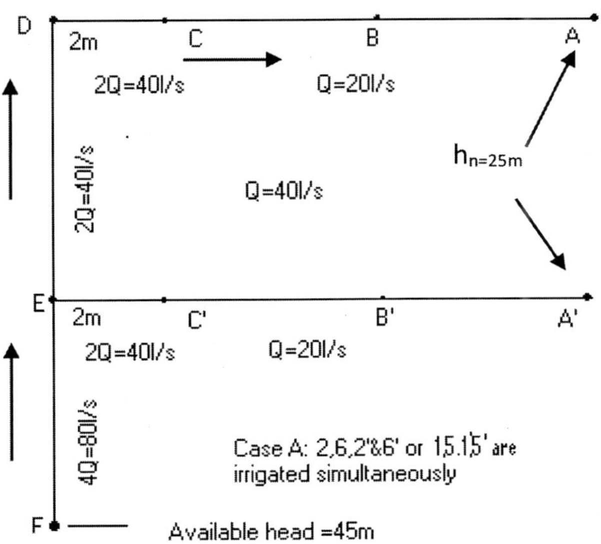

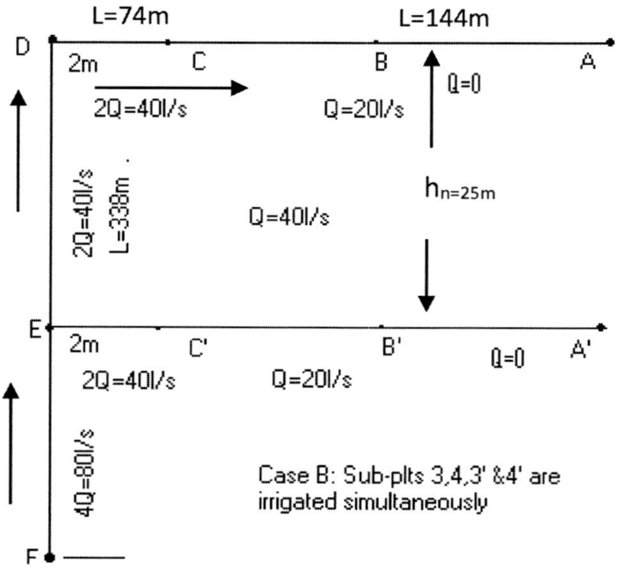

Figure 8.16: Discharges in the mains for cases A& B.

Discharge, 1/s	Section	Length of the section, m	Pipe diameter, mm					
			75	100	125	150	200	250
201/s	AC BC	288 144	-	14.19 7.10	5.04 2.52	2.15 1.08	0.59 0.30	~
401/s	CD DE	74 338	-	13.18 61.11	4.70 21.47	2.01 9.18	0.68 3.11	"
801/s	EF	204	-	-	-	19.96	5.64	1.89

Case B: Design of main B-D-E-F

Using Table 8.14 & Fig. 8.16 the selected diameters for the section BD & DE are 150mm resulting pressure head 15.1 + 25 = 40.1m at E. The remaining head loss for EF = 45 – 40.1 = 4.9m. This states that the section needs 200mm & 250mm pipes. Let x be the length of 200mm diameter pipe.

$$\therefore \frac{5.64}{204}\cdot x+\frac{1.89}{204}(204-x)=4.9$$

$$\text{or, } 0.01838x = 3.01$$

$$\therefore x = 163.74 \approx 164m$$

So, the length of 250 mm pipe = 204 – 164 = 40m

Summary of pipe selection:

Section	Length of pipe, m					
	75mm	100mm	125mm	150mm	200mm	250mm
CD+BC=BD			288	220		
DE				338		
EF					164	40
				558	164	40

Plot II

Case A: Design of main A′-C′-E′-F′

Using Table 8.14 & Fig. 8.16 the selected diameters are 125mm for the section A′C′ & C′E′ resulting pressure head 5.04 + 4.7 + 25 = 34.74m at E. The remaining head loss for EF = 45 – 34.74 = 10.26m. This states that the section needs 150mm & 200mm pipes. Let x be the length of 150mm diameter pipe.

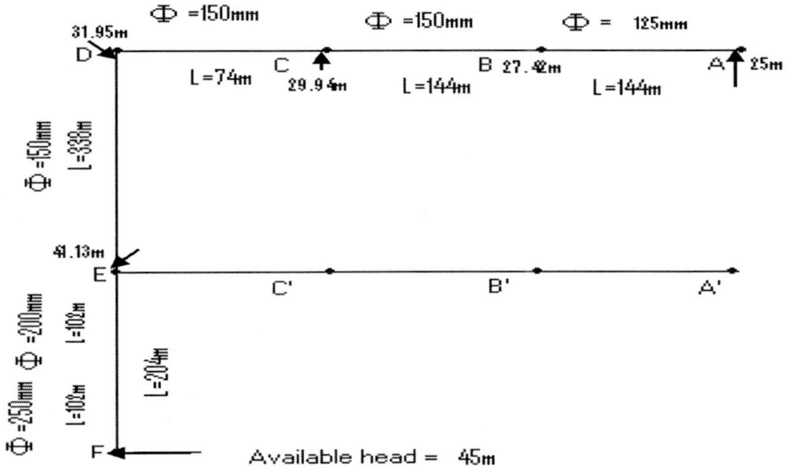

Figure 8.17: Proposed solution in Plot I

$$\therefore \frac{19.96}{204} \cdot x + \frac{5.64}{204}(204 - x) = 10.26$$

$$\text{or, } 0.07x = 4.62$$

$$\therefore x = 66m$$

So, the length of 200mm pipe=204-66=138m

Case B: Design of B′-E′-F′

Using Table 8.14 & Fig. 8.16 the selected diameters are 125mm & 100mm for the section C′E′ & B′C′ resulting pressure head 4.7 + 14.19/2 + 25 = 36.8m at E′. The remaining head loss for EF = 45 – 36.8 = 9.2m. This states that the section needs 150mm & 200mm pipes. Let x be the length of 150mm diameter pipe.

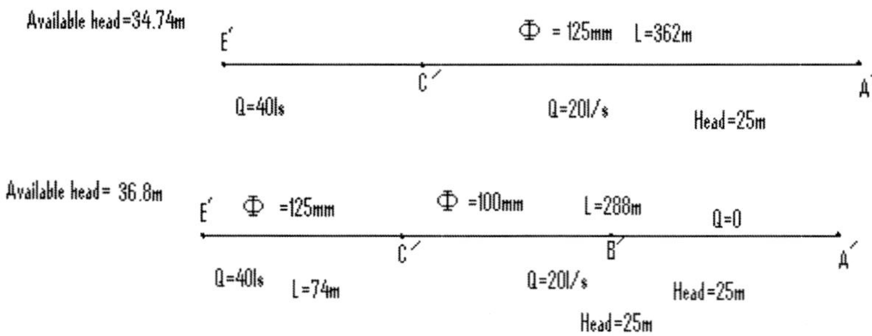

Figure 8.18: Proposed solution in Plot II (Case A & B)

$$\therefore \frac{19.96}{204}\cdot x + \frac{5.64}{204}(204 - x) = 9.2$$

$$\text{or, } 0.07x = 3.56$$

$$\therefore x = 50.85 \approx 51m$$

So, the length of 200mm pipe = 204 – 51 = 153m

In consideration to proposed solutions (Fig. 8.16 & 8.17) it is observed that if 150mm pipe is used in section AB (L = 144m) instead of 125mm a head of 26.02m is available at A which is very close to 25m. Again, this may provide the possibility of avoiding larger diameter 250mm pipe in section EF and cause of higher pressure at C’. However, the value of head at C’ is apparently excessive and therefore suggested to use regulator at the inlets in the manifold in sub-plots 1’ and 2’.

The final solution is stated as below.

Section	**Length of pipe, m**					
	75mm	**100mm**	**125mm**	**150mm**	**200mm**	**250mm**
AB	-	-	-	144	-	-
BD	-	-	-	218	-	-
DE	-	-	-	338	-	-
EF	-	-	-	-	204	-
A’C’	-	-	288	-	-	-
C’E	-			74	-	-
Total	-	-	362	700	204	-

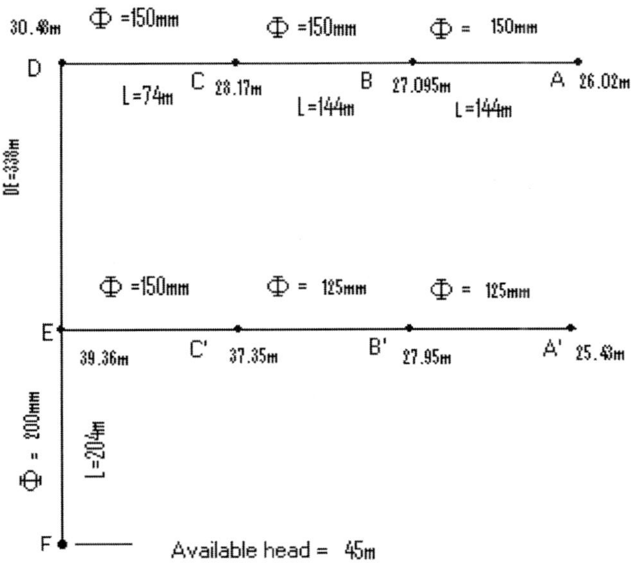

Figure 8.19: Final solution for the mains

The pressure at the inlet of the lateral, $H_n = H_a + 3/4H_f + H_r$

For 7.5cm lateral pipe,

$$H_n = H_a + 3/4H_f + H_r = 22 + 3/4x3.2 + 0.5 = 24.9m$$

For 10.0cm ,, ,, ,

$$H_n = H_a + 3/4H_f + H_r = 22 + 3/4x0.73 + 0.5 = 23.05m$$

Pressure at the farthest end of the lateral, $H_1 = H_a - 1/4H_f + H_r$

For 7.5 cm lateral,

$$H_1 = H_a - 1/4H_f + H_r = 22 - 1/4x3.2 + 0.5 = 21.7m$$

For 10.0 cm ,, ,

$$H_1 = H_a - 1/4H_f + H_r = 22 - 1/4x0.73 + 0.5 = 22.32m$$

Pressure difference between upstream and downstream of the lateral in percent,

For 7.5cm lateral, $$= \frac{24.9 - 21.7}{24.9} x100 = 12.35\%$$

For 10.0cm lateral, $$= \frac{23.05 - 22.32}{23.05} = 3.17\%$$

8.0 Questions & Problems

8.1. Define sprinkler irrigation. What are the advantages and disadvantages of sprinkler irrigation? What are the other uses of it?

8.2. Write the brief history of development of sprinkler irrigation in India.

8.3. Discuss the subsidy schemes to sprinkler irrigation in India. Why the subsidy is provided?

8.4. Classify the sprinkler irrigation system based on portability.

8.5. Describe the components of sprinkler irrigation system with the necessary sketch of it.

8.6. Discuss the inventory of the area for designing the sprinkler irrigation system.

8.7. How the capacity of the sprinkler irrigation system and application rate of individual sprinkler is determined?

8.8. A sprinkler system has 4 laterals each of 100m length with spacing 18m.The sprinklers are at 16m interval on the laterals and apply water at the rate of 0.95cm/h. What is the sprinkler system capacity?

Ans. 18.0l/s

8.9. The following data were obtained from a sprinkler irrigation system.

Sprinkler system = portable

Lateral length = 150m

Spacing of lateral = 15m

Spacing of sprinkler = 15m

No. of laterals = 2

Application rate of sprinklers = 1.25cm/h

Area of the field = 3Ha

Depth of water application = 5cm

Working hours in day = 8

Efficiency of sprinkler system = 80%

What is the time required to irrigate the field if half an hour is allowed in each shifting? What is the system capacity?

Ans. 3.5 days, 23.44l/s

8.10. A solid set sprinkler system with 5 numbers of sprinklers twin nozzle sprinklers of 4mmx3mm diameter operating under 3kg/cm^2 pressure each with coefficient of discharge 0.95. The sprinkler spacing is

12mx15m. The lateral length length and spacing are 120m and 15m respectively. Determine the discharge of each sprinkler, (b) sprinkler system capacity at 80% efficiency.

Ans. (a) 1629.17l/h (b) 28.28l/s

8.11. The sprinklers of 4mm diameter are set at 20m interval on the lateral and each discharges 1.25l/s under operating pressure of 2kg/cm^2. Determine the quality of spray, radius of spray and overlapping.

Ans. $\mathbf{P_d}$ **= 7.28, R = 12.07m, Overlapping = 45.62%.**

8.12. A sprinkler of nozzle diameter 4.5mm discharges 1.50l/s. What is the required pressure in the sprinkler for 10m radius of spray? What is the quality of spray under this pressure?

Ans. R = 12.19m, $\mathbf{P_d}$ **= 4.13**

8.13. Select the appropriate answer from the following.

1. Sprinkler irrigation has started in India during

   a) 1930s　　b) 1940s

   c) 1950s　　d) 1960s

2. The sprinkler nozzle rotates by

   a) 45^0　　b) 90^0

   c) 135^0　　d) 180^0

3. Permanent sprinkler system is suitable to irrigate

   a) Pulses　　b) Vegetables

   c) Field crops　　d) Orchards

4. The sprinkler system is designed for peak consumptive use rate of

   a) 1 day　　2) One week

   c) 10 days　　d) Fortnight

5. Among the following land slopes higher application rate can be suggested for

   a) 0.5%　　b) 2%

   c) 5%　　d) 10%

6. A sprinkler of spacing 20m x 20m sprays at the rate of 0.5l/s for 1 hour. The depth of water application is

   a) 2 mm　　b) 4.5 mm

   c) 5 mm　　d) 5.5 mm

7. The infiltration capacity of a soil is 0.7 cm/h. The sprinkler spacing is 12m x 16m. The application rate of sprinkler is

   a) 0.37l/s　　b) 0.55l/s

   c) 0.63l/s　　d) 0.74l/s

8. A sprinkler nozzle of diameter 4.5mm discharges 0.35l/s. The operating pressure of the sprinkler is

   a) 14.16m　　b) 22.65m

   c) 27.35m　　d) 30.5m

9. A sprinkler nozzle of diameter 0.5cm operates under 3.5kg/cm^2 pressure. The expected radius of the wetted area is

   a) 5.64m　　b) 10.75m

   c) 15.35m　　d) 17.86m

10. A sprinkler discharges at the rate of 0.5l/s under 25m pressure. The quality of spray said to be

    a) Good　　b) Better

    c) Excellent　　d) Wasting of energy

Ans. 1. (c) 2. (b) 3. (d) 4. (a) 5. (a) 6. (b) 7. (a) 8. (c) 9. (d) 10. (d)

8.14. Write True or False of the following.

1. Low pressure in sprinkler the droplets are larger and fall away from the sprinkler.
2. At high pressure the droplets fall near the sprinkler.
3. The depth of water is progressively diminishing towards the sprinkler.
4. In sprinkler irrigation higher overlapping is required for triangular distribution pattern.
5. In designing sprinkler irrigation system undulation and slopes of the field need not to be known.
6. Discharge rate of sprinkler depends on the soil characteristics and land slope.
7. Sprinkler irrigation application rate is higher in sandy loam to clay loam.
8. The root zone depth of cotton is more than apple.
9. The location of the pump is usually fixed with location of source of water in sprinkler irrigation system

10. The point of highest elevation or middle of the field may be the best coice for the location of the pump.
11. The lateral of the sprinkler system follows the steepest slope.
12. The main of the sprinkler system in a slope field follows the level surface.

Ans. 1. True 2.True 3. False 4. False 5. False 6. True 7. True 8. False 9. True 10. True 11. False 12. False

References

Michael, A.M. (1978). Irrigation Principle and Practices. Vikas Publishing House Pvt. Ltd., New Delhi. p.640-647.

Benami, A. & A. Ofen (1984). Irrigation Engineering. Irrigation Engineering Scientific Publications (IESP), (Haifa), Israel with the International Irrigation Information center (IIIC), P.O.B.49, 50250 Bet Dagan, Volcani Centre, Israel: P.O.B.8500, KIG 3H9, Canada.

Hallmark (2004). Hallmark Aquaequipment Pvt. Ltd. 208. Rash Behari Avenue, 2nd Floor, Kolkata.

INCID (1994). Drip Irrigation in India. pp.115.

Rungta (2004). Sprinkler Irrigation System. Electronic complex, Kushalguda, Hyderabad.

Appendix H

Table H-1: Friction loss in meters per 100 meters in lateral line of portable aluminium pipe with couplings (based on Scobey's formula and 9 meters pipe line)

[Adapted from Michael, 1978]

Flow, l/s	Diameter of pipes				
	5.0 cm Ks 0.34	7.5 cm Ks 0.33	10.0 cm Ks 0.32	12.5 cm Ks 0.32	15.0 cm Ks 0.32
1.26	0.32				
1.89	2.53				
2.52	4.49	0.565	0.130		
3.15	6.85	0.858	0.198		
3.79	9.67	1.21	0.280		
4.42	12.9	1.63	0.376	0.122	
5.05	16.7	2.10	0.484	0.157	
5.68	20.8	2.63	0.605	0.196	
6.31	25.4	3.20	0.738	0.240	0.099
7.57		4.54	1.04	0.339	0.140
8.83		6.09	1.40	0.454	0.188
10.10		7.85	.80	0.590	0.242
11.36		9.82	2.26	0.733	0.302
12.62		12.0	2.76	0.896	0.370
13.88		14.4	3.30	1.07	0.443
15.14		16.9	3.90	1.26	0.522
16.41		19.7	4.54	1.47	0.608
17.67		22.8	5.22	1.70	0.700
18.93		25.9	5.96	1.93	0.798
20.19		29.3	6.74	2.18	0.904
21.45		32.8	7.56	2.45	1.02
22.72		36.6	8.40	2.74	1.13
23.98		40.6	9.36	3.03	1.26

25.24		44.7	10.3	3.34	1.38
26.50			11.3	3.66	1.51
27.76			12.3	4.00	1.66
29.03			13.4	4.35	1.80
30.29			14.6	4.72	1.95
31.55			15.8	5.10	2.12
34.70			18.9	6.12	2.52
37.86			22.2	7.22	2.98
41.01			25.9	8.40	3.46
44.17			29.8	9.68	3.99
47.32			33.8	11.0	4.54
50.48				12.5	5.15
53.63				14.0	5.78
56.79				15.6	6.44
59.94				17.3	7.14
63.10				19.0	7.86

Table H-2: Scobey's pipe friction coefficient Ks

[Adapted from Michael, 1978]

Type of pipe material	Ks
Asbestos-cement, cement	0.32
Cement-lined cast iron, dipped and wrapped steel	0.36
Unprotected steel (new), aluminium pipe with couplings	0.40
Slightly used steel	0.44
Fifteen year old steel or iron	0.48
Rough interior	0.54
Very rough interior	0.60

Table H-3: Friction loss in meters per 100 meters in main line of portable aluminium pipe with couplings

Flow, l/s	**Diameter of pipes**						
	7.5 cm	**10.0 cm**	**12.5 cm**	**15.0 cm**	**17.5cm**	**20.0 cm**	**25.0 cm**
2.52	0.658	0.157					
3.15	1.006	0.239					
3.79	1.423	0.339					
4.42	1.606	0.449	0.150				
5.05	2.457	0.548	0.193				
5.68	3.073	0.731	0.242				
6.31	3.754	0.893	0.295	0.120			

7.57	5.307	1.263	0.417	0.170			
8.83	7.113	1.693	0.560	0.227			
10.10	9.169	2.182	0.721	0.293			
11.36	11.47	2.729	0.967	0.366			
12.62	14.01	3.333	1.102	0.448	0.209		
13.88	16.79	3.996	1.321	0.537	0.251		
15.14	19.81	4.713	1.558	0.633	0.296		
16.41	23.06	5.489	1.814	0.737	0.344		
17.67	26.55	6.316	2.089	0.849	0.397		
18.93	30.27	7.203	2.381	0.967	0.452	0.235	
20.19	34.22	8.142	2.092	1.094	0.511	0.265	
21.45	38.39	9.133	3.020	1.227	0.573	0.298	
22.72	42.80	10.18	3.366	1.368	0.639	0.332	
23.98	47.43	11.29	3.731	1.516	0.708	0.368	
25.24	52.28	12.44	4.513	1.671	0.781	0.399	0.136
26.50		13.65	4.930	1.833	0.857	0.445	0.149
27.76		14.57	5.364	1.988	0.936	0.486	0.163
29.03		16.23	5.815	2.179	1.019	0.529	0.177
30.29		17.59	6.284	2.363	1.104	0.573	0.192
31.55		19.01	7.532	2.554	1.193	0.620	0.208
34.70		22.79	8.886	3.060	1.430	0.742	0.249
37.86		26.88	10.35	3.611	1.687	0.976	0.294
41.01		31.30	11.91	4.204	1.965	1.020	0.342
44.17		36.04	13.58	4.839	2.262	1.174	0.394
47.32		41.08	15.35	5.517	2.520	1.339	0.449
50.48			17.22	6.237	2.915	1.513	0.507
53.63			19.20	6.999	3.271	1.698	0.569
56.79			21.28	7.801	3.66	1.893	0.635
59.94			23.45	8.645	4.041	2.097	0.703
63.10			28.11	9.530	4.454	2.312	0.775
69.49			31.75	11.42	5.338	2.771	0.929
75.72				13.58	6.298	3.269	1.096
82.03				15.69	7.333	3.886	1.277
88.34				18.06	8.441	4.382	1.470
94.65				20.59	9.624	4.996	1.675
101.0				23.28	10.88	5.548	1.894
107.3				26.12	12.21	6.337	2.125
114.0					13.63	7.064	2.369

120.0					15.08	7.829	2.625
126.0					16.62	8.630	

Note: Where 6.1m sections of pipes are used, increase values shown in Table by 7.0 percent. Where 12.2m sections of pipes are used, decrease values shown in the Table by 3.0 percent.

Table H-4: Friction he3ad loss in semi-rigid plastic irrigation pipelines manufactured of PVC or asbestos compounds

Flow, l/s	Diameter of pipe						
	2.5 cm	**3.18 cm**	**3.8 cm**	**5.08 cm**	**6.25 cm**	**7.61 cm**	**8.9 cm**
0.160	0.15	0.04	0.02				
0.252	0.54	0.17	0.09	0.03	0.01		
0.378	1.15	0.37	0.19	0.06	0.02		
0.503	1.97	0.63	0.32	0.11	0.04	0.01	
0.631	2.98	0.95	0.49	0.16	0.06	0.02	0.01
0.946	6.32	2.03	1.04	0.35	0.14	0.05	0.02
1.26	10.79	3.46	1.78	0.60	0.23	0.09	0.04
1.58	16.30	5.22	2.70	0.91	0.36	0.12	0.07
1.89	22.86	7.32	3.78	1.27	0.50	0.19	0.10
2.21		9.75	5.03	1.70	0.67	0.25	0.13
2.52		12.46	6.46	2.18	0.86	0.32	0.17
2.84		15.51	8.02	2.71	1.07	0.40	0.21
3.15		18.87	9.75	3.30	1.30	0.49	0.25
3.47		22.48	11.64	3.94	1.54	0.59	0.30
3.89			13.64	4.62	1.81	0.69	0.36
4.10			15.85	5.36	2.10	0.80	0.41
4.42			18.19	6.14	2.42	0.92	0.47
4.73			20.65	6.99	2.75	1.06	0.55
5.05			23.28	7.86	3.10	1.19	0.62
5.36				8.81	3.47	1.33	0.69
5.68				9.79	3.85	1.48	0.77
5.99				10.82	4.25	1.64	0.85
6.31				11.89	4.69	1.80	0.93
6.94				14.21	5.59	2.14	1.11
7.59				16.69	6.56	2.52	1.31
8.20				16.35	7.63	2.92	1.53
8.83				22.21	8.73	3.36	1.75
9.46					9.94	3.82	1.99
10.10					11.20	4.29	2.24
10.73					12.51	4.80	2.50

11.36					13.90	5.35	2.79
11.99					15.39	5.93	3.08
12.62					16.9*1	6.50	3.38
13.88					20.19	7.77	4.04
15.14					23.73	9.12	4.76
16.41						10.57	5.51
17.67						12.11	6.32
18.9*3						13.78	7.18
20.16						15.52	8.10
21.48						17.37	9.07
22.72						19.27	10.08
23.98						21.33	11.13
25.24						23.45	12.22
26.50							13.40
27.76							14.59
29.03							15.86
30.29							17.15
31.55							17.50

Note: Table based on Hazen William's equation, C=150

To find friction head loss in PVC or asbestos pipe having a standard dimension ratio other than 21, the values in the Table should be multiplied by the appropriate conversion factor shown below:

SDR No.	**Conversion factor**
13.5	1.35
17	1.13
21	1.00
26	0.91
32.5	0.84

Table H-5: Friction head loss in semi-rigid plastic irrigation pipelines manufactured of PVC or asbestos compounds

(Standard dimension ratio, SDR 21)[Adapted from Michael, 1978]

Flow, l/s	**Diameter of pipe**					
	10.0 cm	**12.5 cm**	**15.0 cm**	**20.0 cm**	**25.0 cm**	**30.0 cm**
0.946	0.01					
1.26	0.02					
1.58	0.04	0.01				
1.89	0.05	0.02				

2.21	0.07	0.02	0.01			
2.52	0.09	0.03	0.01			
2.84	0.12	0.04	0.01			
3.15	0.14	0.05	0.02			
3.47	0.17	0.06	0.02			
3.79	0.20	0.07	0.03			
4.10	0.23	0.08	0.03	0.01		
4.42	0.27	0.09	0.04	0.01		
4.73	0.31	0.11	0.04	0.01		
5.03	0.35	0.12	0.05	0.01		
5.36	0.39	0.14	0.05	0.01		
5.68	0.43	0.15	0.06	0.01		
5.99	0.48	0.17	0.07	0.02		
6.31	0.52	0.19	0.07	0.02		
6.84	0.63	0.22	0.09	0.02		
7.57	0.74	0.26	0.10	0.03	0.01	
8.20	0.85	0.30	0.12	0.03	0.01	
8.83	0.98	0.35	0.14	0.04	0.01	
9.46	1.11	0.40	0.16	0.05	0.01	
10.10	1.26	0.44	0.19	0.05	0.01	
10.73	1.41	0.49	0.21	0.06	0.02	
11.36	1.57	0.55	0.24	0.07	0.02	0.01
11.99	1.73	0.61	0.26	0.07	0.02	0.01
12.62	1.90	0.67	0.29	0.08	0.02	0.01
13.88	2.28	0.81	0.34	0.09	0.03	0.01
15.14	2.67	0.95	0.40	0.10	0.03	0.01
16.41	3.10	1.10	0.46	0.12	0.04	0.02
17.67	3.56	1.26	0.54	0.14	0.05	0.02
18.93	4.04	1.43	0.61	0.17	0.05	0.02
20.19	4.56	1.62	0.69	0.19	0.06	0.03
21.45	5.10	1.82	0.77	0.21	0.07	0.03
22.72	5.67	2.02	0.86	0.24	0.08	0.03
23.98	6.26	2.22	0.95	0.26	0.09	0.04
25.24	6.90	2.45	1.04	0.28	0.10	0.04
26.50	7.55	2.69	1.14	0.31	0.10	0.05
27.76	8.23	2.92	1.25	0.34	0.11	0.05
29.03	8.94	3.18	1.35	0.37	0.12	0.06
30.29	9.67	3.44	1.46	0.41	0.14	0.06

31.55	10.42	3.70	1.58	0.43	0.15	0.06
34.70	12.44	4.42	1.89	0.52	0.18	0.07
37.86	14.61	5.21	2.22	0.61	0.21	0.09

SDR No.	**Conversion factor**
13.5	1.35
17	1.13
21	1.00
26	0.91
32.5	0.84